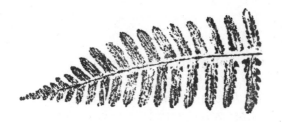

COMING

EARTH
CHANGES

By the same author

A Psychic Interpretation of Some Late-Cenozoic Events Compared with Selected Scientific Data (1959)

Earth Changes: Past—Present—Future (1960)

"New Portrait of Our Planet" and the Cayce Predictions (1961)

Updating Earth Changes (1964)

Atlantis at Bimini? (1968)

Earth Changes: Past—Present—Future (1980) (By "A Geologist," Part II of a book by Hugh Lynn Cayce, entitled *Earth Changes Update*)

Earth Changes Update: Searching for Truth Within Psychic Visions and Scientific Insight. A bimonthly newsletter largely directed to updating the information and interpretations found in this book, 1997—present. Published by A.R.E. Press.

COMING
EARTH
CHANGES

Causes and Consequences of the Approaching Pole Shift

by William Hutton

ARE PRESS

ASSOCIATION FOR RESEARCH AND ENLIGHTENMENT

A.R.E. Press • Virginia Beach • Virginia

A.R.E. Press
215 67th Street
Virginia Beach, VA 23451-2061

Library of Congress Cataloging-in-Publication Data
    Hutton, William, 1931-
    Coming earth changes : causes and consequences of the
approaching pole shift / by William Hutton.
        p.   cm.
    Rev. ed. of: Earth changes update / Hugh Lynn Cayce.
©1980.
    ISBN 0-87604-361-9
    1. Catastrophes (Geology). 2. Geophysics. 3. Cayce, Edgar,
1877-1945. 4. Metaphysics. 5. Atlantis. I. Cayce, Hugh Lynn.
Earth changes update. II. Title.
QE506.H94   1996
133.3-dc20   96-15715

Cover design by Lightbourne Images

Contents

Foreword

The clairvoyant predictions of Edgar Cayce about future changes in planet Earth are, from our myopic human vantage point, nothing less than sensational. Volcanoes erupting, continents rising from beneath the sea, Earth quaking, coastal areas flooding, poles shifting, great cities collapsing—all this and more is in the scenario described in this remarkable man's psychic readings. But what are we to make of it?

When I first learned about Edgar Cayce in the 1960s, it was these predictions that most piqued my interest. As a reporter, working for newspapers on the West Coast of the United States, I was fascinated by Cayce's predictions of the destruction of San Francisco and Los Angeles in the near future. Even to a skeptical journalist this was believable. The possibility of such a major story hooked me.

Ever since then I have been intrigued with the forecasts, visions, and philosophical wisdom that poured from Edgar Cayce when he went into a hypnotic trance. In the 1980s, however, I encountered an entirely different perspective. Everett Irion, a student of the Cayce material, thought the descriptions of Earth changes should be read as an allegory, such as *Pilgrim's Progress*. Cayce's dire forecasts, he believed, were a way of revealing spiritual meaning through concrete or material forms. They were not to be taken literally, but symbolically.

Since my own life was being rocked by unexpected eruptions and quakes, I wondered if this was what Irion had alluded to. The same thing was happening to others I knew who were consciously following a spiritual path. The meaning seemed clear: Life is ever changing, and seemingly innocent souls are often shaken to their foundation by cataclysmic personal events that challenge our faith in a Higher Power. In attempting to understand and come to terms with the nature of life, I felt that we had either to believe that the Universe was a hostile, uncaring environment or that "the Universe cared" and that we must learn to trust God's process. I chose the latter option. Although it often baffles me in any given moment, I am learning to trust that process.

While this progression in my spiritual philosophy allowed me to take Irion's perspective seriously, I thought the possibility remains that the poles may shift and Mt. Vesuvius could still blast off and trigger a chain of events that changes the face of the planet as we know it. Indeed, we may be in for planetary as well as personal changes.

Still another perspective emerged and gained popularity within the Association for Research and Enlightenment, the organization Cayce founded. Cayce had said that physical changes in the earth were affected by the spirit and attitude of its inhabitants. That is, we are all responsible for what happens in the world and to the planet. Enough people with a positive, loving attitude can save the world. The biblical story of Jonah supports this idea, for the city of Nineveh was said to be threatened with destruction because of the hate and selfishness of its inhabitants.

Although this concept may be difficult for our scientifically oriented culture to accept, it has broad appeal in spiritual circles where the power of prayer and meditation to alter the course of events is widely respected. Even some scientists are beginning to embrace such metaphysical beliefs.

One of them is the author of this engaging book, *Coming Earth Changes*, which examines the possibility of Earth changes from both a scientific and a spiritual vantage point. Writing under a pen name, William Hutton, he is a geologist who has taught at three universities, written scores of research papers, and worked as a director of geoscience research programs funded by the United States government. He has also written extensively on this topic and been a member of the A.R.E. for some forty years.

In his analysis of the possibility of major Earth changes, Hutton has accumulated revealing evidence from field reports of colleagues who specialize in tectonics and have made in-depth studies of what is happening beneath the surface of the Earth—tracking the crust's movements in response to the buildup of molten material contained beneath the crust. What we realize dramatically from his account is that Earth is constantly changing, and has been since creation. Indeed, data compiled by the scientific community tend to lend credence to Cayce's forecasts of change, including damage to some of California's great and admittedly vulnerable cities, and to Japan and many other areas as well.

Hutton approaches his task with what I think of as the truly scientific method, an open-minded willingness to examine Cayce's ideas, test them against what science has thus far learned, and draw his own independent conclusions. Scientific theories are constantly changing, of course, as new evidence is gathered. So scientific truth is only as good as the information on which it is based, and new data are constantly coming in. Thus Hutton's account is replete with up-to-the-minute information. In some instances, he notes, it took science decades to catch up with Cayce and to verify, scientifically, what he had revealed psychically years before.

To further test Cayce's reliability, Hutton dug into scientific archives to see if there is any correlation between events that Cayce predicted would occur in past years and the actual record of geophysical events. Hutton's findings prove quite impressive and lead to one conclusion: Edgar Cayce cannot be dismissed, nor can his predictions.

Finally, as a spiritual seeker himself, Hutton is open to the possibility that we can yet "save the world" by spiritual means.

Whether one treats *Coming Earth Changes* as an allegory, or as a wake-up call for a spiritual revival, or as an overview of how planet Earth may be transformed in the new millennium, is for each reader to determine. But surely these changes will affect us all. Our challenge is to respond to them with the best that is in us, for the sake of ourselves, our families, and our planet.

A. Robert Smith
Editor
Venture Inward

Introduction

I first heard of Edgar Cayce's readings on ancient man, lost continents, and the building of the Great Pyramid when I was a graduate student in geology. Being full of the latest Earth-science knowledge, I thought I was qualified to critique these readings in a magazine article. I contacted Hugh Lynn Cayce, eldest son of the then deceased Edgar Cayce, about my idea. He thought it worthwhile and sent me a number of his father's readings on ancient continents and future Earth changes.

As I perused the Cayce material, I realized that many of the assertions made as to the existence of Atlantis and Lemuria, and to the metaphysical relationship of societal behavior and resultant Earth changes, would be hard to corroborate. But some of the readings sounded plausible, and so I began to develop a scientific database for the readings that I felt might someday be verifiable or capable of being clearly disproved.

One of the first readings that I studied stated that what is now the Sahara Desert was once an inhabited and very fertile land and that "the Nile entered into the Atlantic Ocean" (364-13). And a fragment of reading 5748-1 read, " . . . .the waters then entering the now Atlantic from the Nile region rather than flowing northward." My reaction to these statements about the Nile was, "What a preposterous thing to say!"

But about 30 years later, I came across an article in *Science* magazine that described a new hypothesis, based on satellite imagery studies. It stated:

> "A river of Amazonian proportions once spanned the African continent. It flowed from near the present-day Red Sea, across the path of today's Nile, through the eastern Sahara Desert . . . across central Africa with its [now] towering mountains, and into the Atlantic 4,500 kilometers from the system's head-waters.[1]

According to this scientific hypothesis, the river flowed into the Atlantic for 10 to 20 million years, until around 15 million years ago when Earth changes began slowly to divide the drainage system into several basins. As for the "[now] towering mountains" mentioned in *Science*, the following Cayce reading makes a similar observation:

> **In the one [incarnation] before this we find again in this same land now called Egypt (this before the mountains rose in the south, and when the**

waters called the Nile then emptied into what is *now* the Atlantic ocean.)

276-2
February 20, 1931

The scope of the Edgar Cayce readings on what loosely may be called "Earth changes" is sufficiently great that the mind is incapable of remembering most of the changes in any detail. Instead, we tend to carry around in our heads only short phrases of readings that together make up our particular set of thoughts on the subject. Many of the best-remembered phrases relate to potential *future* geophysical events, especially those forecast to begin to occur around the present time or to the end of the century.

We begin with a brief review of the accuracy of the readings' predictions of forthcoming Earth changes. This review will consider Earth-change predictions that some previous commentators on the readings have said were not fulfilled. We'll discuss such negative assessments in terms of related readings that take a more broadly focused view and in light of recent geophysical data.

Next, we will consider the predictions of pole shift for the period 1998-2001 A.D., because such a pole shift would be the driving force behind most of the other Earth changes mentioned in the readings. An analysis of current geophysical events in the Mediterranean and South Pacific regions will then be presented, events that coincide with the readings' contention that the prophesied Earth changes have indeed begun. We'll then consider trends in seismotectonic events in the western U.S., like the Northridge, California, earthquake of January, 1994, and examine what might occur in many other places along the U.S. west coast if certain predictions of destructive tectonic events are realized. (Some events were said to be inevitable.) This approach to the Earth-change readings will lead us to a review of scientific forecasts for potentially destructive geophysical events in other parts of the U.S. and the world. We will find that these scientific forecasts tend to coincide with many of the Earth-change predictions in the Cayce readings.

Some of my colleagues in the scientific community are sure to think that I've been too selective in the choice of scientific materials I've used to support my interpretations of Edgar Cayce's Earth-change readings. I can hear them ask: How can you justify this approach? Who will believe you? There's no evidence for Atlantis or Lemuria.

Progress in understanding can often benefit from a biased approach. Indeed, the great Princeton geologist, M.F. Buddington, is said to have commented, "No progress without prejudice!" Jogging the logical, science-based side of one's mind with unfamiliar psychic information may provide an insight here and there that can prove useful later on. Presumably, starting in 1998, we will be increasingly required to know the world around us by means of a synthesis of left- and right-brained cognition.

If any scientifically oriented readers find it hard to accept my presentation, I think that they and others may benefit from the expanded understanding of life found in the readings, as I have. And if and when any of the specific Earth-change readings are actualized, there is plenty of additional material in this book to transport anyone to expanded levels of understanding. Human beings are their understanding, not their heap of facts and inventions.

For readers unfamiliar with the background of Edgar Cayce, he grew up in rural Kentucky a century ago and began to show evidence of remarkable psychic ability as a young man. He found that when he was under hypnosis, he could answer questions on subjects he had never studied, including diagnosing illnesses that baffled medical doctors. The questions he answered while in a sleeplike state were all recorded by his secretary and transcribed as "readings." The subject matter of the more than 14,000 readings on file at the Edgar Cayce Foundation in Virginia Beach, Virginia, covers a broad spectrum from human-health to cosmic metaphysical concepts, from ancient history to future predictions, including those on changes to come in our planet.

To give you an idea of the appearance of an Edgar Cayce reading, a few paragraphs of one are presented. The reading number and the date when given will appear at the end of each reading quoted in this book. The complete readings, including backup correspondence and cross references to other readings, are available to the public on CD-ROMs for either PC or Macintosh operating systems.[2]

This psychic reading was given by Edgar Cayce at his office, 332 Grafton Avenue, Dayton, Ohio, May 28, 1925, in accordance with request made by self—Edgar Cayce, at 11:30 A.M., D.S.T. Present: Edgar Cayce; Gertrude Cayce (conductor); Gladys Davis (stenographer).

Mrs. Cayce: You will have before you the psychic work of Edgar Cayce, present in this room, especially that phase pertaining to life readings and former appearances of individuals in the Earth's plane. In several readings there has been given information concerning the second ruler in Egypt who gave the first laws concerning man's relation to the Higher Forces. You will give me an outline of this teaching and how same was given to the people.

Mr. Cayce: Yes, we have the work here and that phase concerning the indwelling in the Earth's plane of those who first gave laws concerning indwelling of Higher Forces in man. In giving such in an understandable manner to man of today, [it is] necessary that the conditions of the Earth's surface and the position of man in the Earth's plane be understood, for the change has come often since this period, era, age, of man's Earthly indwelling, for then at that period, only the lands now known as the Sahara and the Nile region appeared on the now African shores; that in Tibet, Mongolia, Caucasia and Norway [appeared] in Asia and Europe; that in the southern cordilleras and Peru in the southwestern hemisphere, and the plane of now [present] Utah, Arizona, Mexico of the northwestern hemisphere, and the spheres were then in the latitudes much as are presented at the present time.

... man's indwelling [was] then in the Sahara and the upper Nile regions, the waters then entering the now Atlantic from the Nile region rather than flowing northward. The waters in the Tibet and Caucasian entering the North Sea; those in Mongolia entering the South Seas, those in the cordilleras entering the Pacific, those in the plateau entering the Northern Sea.

When the Earth brought forth the seed in her season, and man came in the Earth plane as the lord of that in that sphere, man appeared in five places then at once—the five senses, the five reasons, the five spheres, the five developments, the five nations.

5748-1
March 28, 1925

Now, as we see, as given, how and what the classifications were of the physical in the Earth's plane at that period, the numbers then of human souls in the Earth plane being a hundred and thirty and three million (133,000,000) souls. The beginning then of the understanding of laws as applied from man's viewpoint being in this second rule in the country now Egypt. The rule covering the period of a hundred and ninety and nine (199) years, and the entity giving the chance to the peoples for the study being in the twenty and eighth (28th) year, when [he] began to gather the peoples together for this and surrounding himself with those of that land and of the various lands wherein the human life dwelled at that period. The numbers of the people that came together for the purpose then numbering some forty and four (44).

The Courts as were made were in the tents and the caves of the dwellers of the then chosen priest from the Arabian or Tibetan country, who came as one among those to assist with the astrologer and the soothsayers of the desert of now the eastern and western worlds, and with this the conclave was held for many, many, moons. The period in the world's existence from the present time being ten and one-half million (10,500,000) years, and [in] the changes that have come in the Earth's plane many have risen in the lands. Many lands have disappeared, many have appeared and disappeared again and again during these periods, gradually changing as the condition became to the relative position of the Earth with the other spheres through which man passes in this solar system.

5748-2
May 28, 1925

Readers who are new to Cayce's readings may wonder about their unusual language, their accuracy, and the sources of the information obtained by Cayce. These subjects are addressed in the Appendix.

Readers unfamiliar with the Cayce philosophy should also know that reincarnation is a key to understanding the story of mankind that emerges from his readings; that is, each of us souls has lived many lives here on Earth, experiencing materiality in the process of our development. This explains the reference to "the one [incarnation] before this" in the reading just quoted. The following extract addresses the degree of emphasis to be placed on reincarnation, as expressed in the "life" readings that Cayce gave for various individuals. The reading was given for a woman who planned to construct a new home in Virginia Beach, Virginia, location of the Association for Research and Enlightenment (A.R.E.).

Q Just what will be the nature of my work in my new home?

A That as indicated in the suggestions as to what the body should do in its activities in the present; and it *should* be present in the erection of the quarters for the body!

Q Is it to spread the message of reincarnation as interpreted through the life readings?

A That is merely a part. Remember, the real purpose as should be for each soul is the message of the love of the Savior, Jesus the Christ, for the children of men. That phase of Christian experience (reincarnation) is questioned by many, yet there is this period when the fact needs stressing to answer many questions. But that this is to be the primary fact—reincarnation, no. That is

merely the plan as He demonstrated.

1152-12
September 9, 1941

A few of the topics addressed here were originally presented in the A.R.E. magazine *Venture Inward*, under the pen name of The Geologist. They have been revised and expanded in this comprehensive treatment of the subject and compared with the latest scientific findings from around the world.

I am indebted to A. Robert Smith for his splendid job in editing the manuscript. Rather than merely follow the carpenter's adage, "Cut to fit, beat into place," Bob Smith not only skillfully reorganized the text but also, by his thoughtful questions, coaxed from me the many revisions to the prose that the manuscript so sorely needed.

William Hutton
July 1996

PART ONE
THE PREDICTIONS

1

Cayce's Psychic Visions

Edgar Cayce's psychic visions of coming Earth changes were global in reach, monumental in scope, breathtaking in their implications, and phenomenal in that some of them preceded corroborating scientific discoveries by several decades.

Here are the highlights of the predictions that came through Cayce:

• sinking of the "greater portion of Japan" into the sea (3976-15)

• changing of the upper portion of Europe "in the twinkling of an eye" (3976-15)

• disappearance of the main portion of New York City (1152-11)

- emptying of the Great Lakes into the Gulf of Mexico (1152-11)
- noticeable physical changes in "New York . . . Connecticut, and the like" (311-8)
- disappearance of "the southern portions of Carolina, Georgia," sooner than "New York City itself." (1152-11)
- a shifting of the poles of the Earth in the period 1998 to 2001 (378-16, 826-8, and 3976-15)
- breaking up of the Earth in the western portion of America (3976-15)
- inundations by earthquakes along the southern coast of California and between Salt Lake and southern Nevada following "greater activities in the Vesuvius or Pelée" (270-35)
- disturbances in portions of the west coast, east coast, and central portions of America (1152-11)
- appearance of open waters off northern Greenland (3976-15)
- breaking up of "conditions in the . . . South Pacific" (311-8)
- appearance of new lands off the Caribbean Sea (3976-15)
- shaking of South America "from the uppermost portion to the end" (3976-15)
- appearance of land off the east coast of America (3976-15)
- appearance of land "in the Antarctic off of Tierra del Fuego" (3976-15)
- upheavals in the Arctic and Antarctic that will cause volcanic eruptions in the Torrid areas (3976-15)
- rising from the sea of a temple of Atlantis (378-16)
- rising of Poseidia, "among the first portions of Atlantis to rise . . . from the sea" (958-3)
- many disturbances "in other lands" beyond America (1152-11)

- sinking or rising of the crust in the Mediterranean (Etna) area (311-8)
- sinking of many of the World War II battlefields to become oceans, seas, or bays (1152-11)
- sinking of the coast lines of "many a land" (1152-11)
- appearance of lands in the Atlantic and Pacific oceans (1152-11)

Before commencing our analysis of the possibility of these events happening, we need to consider the accuracy of psychic predictions. No one who makes predictions, be they based on normal scientific data or paranormal intuitive information, is always correct. And so the question of Cayce's accuracy cannot be ignored. His record, so far as I have been able to evaluate it, is mixed. Some predictions were right on target, some apparently missed. Here are a few examples:

What is believed to be the first of geophysical predictions in a Cayce reading were given for a man who was interested in mid- to long-range weather forecasts and their effects on wheat and corn prices. His objective was to make money in the grain-futures markets. The passage quoted below was the last paragraph in the man's reading of August 27, 1926, and came after he had indicated to the entranced Cayce that all of his questions had been answered. In the context of the entire reading, the dates of October 15, 19, and 20, 1926, come up as days when wheat and corn prices would be higher on the London grain market.

> As for the weather conditions, and the effect same will produce on various portions of the earth's sphere, and this in its relation to the conditions in ... the affairs of the world appear the strongest on or about October 15th to 20th—when there

may be expected in the minds, the actions—not only of individuals but in various quarters of the globe, destructive conditions as well as building. In the affairs of man many conditions will arise that will be very, very, strange to the world at present—in religion, in politics, in the moral conditions, and in the attempt to curb or to change such, see? For there will be set in motion [that indicating] when prohibition will be lost in America, see? Violent wind storms—two earthquakes, one occurring in California, another in Japan—tidal waves following, one to the southern portion of the isles near Japan.

195-32
August 27, 1926

Five days later, in a follow-up reading, Cayce was asked:

Q In reading given on August 27th regarding weather, where in California will earthquake predicted October 15th-20th be the worst? Will there be a tidal wave at that period, or where in California?

A Tidal wave being, as is given, in the Far East, the earthquake being in lower California, see?

195-33
September 1, 1926

By researching records for the period covered by that 1926 reading for the commodity investor, I found that the violent wind storms predicted in the reading certainly did occur within the approximate timeframe specified ("*on or about* October 15th to 20th"). According to the *Monthly Weather Review* for October 1926,[1] "October was an exceptionally stormy month and the number of days with gales was considerably above the normal over the greater part of the [North Atlantic] ocean. Several tropical disturbances occurred during the month, three of which were of slight intensity, but the storm that created such havoc in Cuba on the 20th was one of the most severe on record." In the vicinity of the Kuril Islands, "the westerly winds increased to hurricane force on the 14th and 15th . . . " Reports from ships in the vicinity of the Philippine Islands "indicate three and probably four violent storms [typhoons] during the early part of October 1926." And the *India Weather Review* reported "a moderate storm occurred the 15th to 18th of October in the Andaman Sea."[2] All told, this was an impressively accurate prediction of the occurrence of "violent wind storms" for the time frame specified in the reading.

With respect to the earthquake predictions, an earthquake struck California on October 22, 1926. It was composed of two magnitude-6 shocks located just offshore of Monterey at about 36.58° N, 122.2° W.[3] This quake's epicenter was located slightly south of an east-west line dividing California in two across its middle. Technically speaking then, it did occur in *lower* California, as the reading said it would. (Note that in the context of readings 195-32 and -33 above, Lower [capital "L"] California was not meant. And, if Cayce had meant Lower California, he would most likely have said "Baja California" or "California in Mexico.")

The October 22 quake was "perceptible over probably 100,000 square miles"[4] and an isoseismal map of earthquake intensities[5] shows that the area bounded by a line from Half Moon Bay east to Hayward, southward to San Jose and to King City, and then westward to the ocean was subjected to Intensity VI+ on the Modified Mercalli scale. Such an intensity is "felt by all" in the area and may overthrow unstable objects, move heavy furniture, and cause slight damage to plaster and chimneys. An area experiencing Intensity V effects extended eastward of the Intensity VI contour and was bounded on the north and east by an isoseismal line running from Fallon to Vacaville to Stockton to Turlock and on into south-central lower California.

The previous strong California shock to the one of October 22, 1926, occurred on July 25, 1926, and the following one was on January 1, 1927. Thus, the October 22 earthquake occurred within the time window specified by the reading; that is, "on or *about* October 15th to 20th."

The strongest shock in Japan, occurring on or about the 15th to 20th of October 1926 was recorded on October 19. This was a magnitude 6 quake,[3] as calibrated by the Japanese Meteorological Agency. It was an undersea shock whose epicenter was just off the southern tip of Hokkaido at 41.666° N, 143.006° E. Considering that "some experts now view earthquakes as a classic example of a chaotic system" and that many of the nation's leading seismologists now think that earthquakes are "inherently unpredictable,"[6] the Cayce reading did a very respectable job of predicting the general time and place for each of the quakes in California and Japan. Some have suggested that the two quakes that occurred were too insignificant even to have been mentioned, and that the quake predictions therefore failed. But this seems to be a gratuitous argument. The passage quoted from reading 195-32 was meant to be a helpful psychic afterthought advanced to emphasize the turbulent nature of human psychology and mentation, as well as the Earth's geophysical environment, that would affect the grain markets. The purpose was not to identify *significant* earthquakes, as some have conjectured.

No records of a tidal wave, in the sense of a tsunami (seismic sea wave), have been found for the period following the quake in Japan. The undersea location of the epicenter of the Japanese quake, however, meets the first prerequisite for generation of a tsunami. If a tsunami did form, it may have been too small to have been deemed worthy of being recorded in Heck's "List of Seismic Sea Waves."[7] Had such a small tsunami occurred, however, it could have affected the southern portions of Yezo (now Hokkaido) and associated smaller islands in the Kuriles near Japan. According to tsunami records in the International Tsunami Information Center, a small tsunami wave from somewhere in the western Pacific was recorded in Hawaii on October 24, 1926. In spite of the foregoing speculation it is quite possible that, for reasons unknown, the tidal-wave prediction, in the sense of a tsunami-type "tidal wave," simply failed to materialize. Or perhaps the tidal waves envisioned were storm surges due to atmospheric disturbances.

The next geophysical prediction to appear in a Cayce reading was in 1928. It was for a man who lived near Atlanta, Georgia, who had suffered for five years from the aftereffects of a bout of the flu. His reading advised him to get a "change of climate, change of scene, change of actions for the body . . . " Cayce was asked:

> Q Would the climate of San Diego, California, be beneficial to him and his family?
> A That ABOVE San Diego would be more beneficial than that near to San Diego. Beware of the quakes as will occur there a little later on.
> Q How soon will they occur?
> A Eighteen months to two years.
>
> 4283-5
> September 15, 1928

Here again, Cayce had not been asked to forecast future events, but he apparently felt it only fair to warn his client of possible alarming earthquake conditions he would encounter if he moved to the San Diego area. To determine how accurate was his forecast, I researched historical records at the National Earthquake Information Center, maintained by the U.S. Geological Service in Golden, Colorado, for the six-month timeframe— March 1 through September 30, 1930—that was specified ("eighteen months to two years" from the date of that reading). The area around San Diego during this target period was shaken by a series of six minor quakes north or northeast of San Diego. None occurred within the city limits. The largest of the quakes occurred on May 12, 1930. There was a magnitude 4.0 ("felt" intensity) quake just east of Julian, about 35 miles northeast of San Diego, followed 13 hours later by a magnitude 4.2 (Intensity IV) quake near Ramona, some 20 miles north-northeast of the city limits.

While these were mild to a quake-familiar native Californian, they could have been disconcerting to a Georgian and his family, newly arrived in California. Perhaps Edgar Cayce, who had given this man five readings over a five-year period, felt it necessary to warn him of the quakes so that he would not feel let down by his psychic friend shortly after moving to the West Coast.

Was the quake prediction by the readings an accurate one? Definitely so. Moreover, Cayce's ability to pinpoint the approximate time of the 1930 quake is all the more impressive when you realize that earthquakes of this "felt" intensity are not everyday events, even in California. The next one that appeared was more than two years later—a magnitude 4 quake on June 23, 1932. In other words, there were only two quakes of this intensity in nearly four years following that reading, but when Cayce was asked when to expect one of these events, he was on target in predicting one would occur within "Eighteen months to two years."

On the other hand, some events predicted in the Cayce readings seem not to have occurred at their appointed times. In 1932, for example, Cayce was asked:

Q Are there to be physical changes in the Earth's surface in Alabama?
A Not for some period yet.
Q When will the changes begin?
A Thirty-six to thirty-eight.
Q What part of the State will be affected?
A The northwestern part, and the extreme southwestern part.

311-9
August 6, 1932

Q Are the physical changes in Alabama predicted for 1936-38 to be gradual or sudden changes?
A Gradual.
Q What form will they take?
A To be sure, that may depend upon much that deals with the metaphysical, as well as to that people called actual or in truth! for, as understood—or should be understood by the entity—there are those conditions that in the activity of individuals, in line of thought and endeavor, keep oft many a city and many a land intact through their application of the spiritual laws in their associations with individuals. This will take more of the form here in the change, as we find, through the sinking of portions with the following up of the inundations by this overflow.

311-10
November 19, 1932

There is no obvious evidence that the predicted changes took place in 1936-38. On the other hand, gradual changes may have *begun* in 1936 to 1938 and could even now be manifesting there. But such changes may be too subtle or too slow to detect as yet. At the same time, it is possible to propose that, as Edgar Cayce's son, Hugh Lynn Cayce, wrote, "Apparently . . . there were some very positive attitudes developed in Alabama which mitigated the changes."[9] That is, in the metaphysical logic of the readings, proper application of spiritual laws in Alabama could have negated the predicted surficial changes there.

The Alabama and Tombigbee Rivers follow an ancient, elongated crustal unit, bounded by faults along its sides, called the Mobile Graben. (See Fig. 1.) Also, the (interpreted) location of the Florida-Bahamas Transfer Fault runs through southwestern Alabama, as does the Gulf Rim Fault Zone (also called the Regional Peripheral Fault Trend). Reactivation along any of these structural features, especially the Mobile Graben, could lead to subsidence in southwestern Alabama.

Perhaps the easiest way to attempt verification of sinking of the extreme southwestern part of Alabama would be to compare the U.S. Soil Conservation Service's air photos of the region around Mobile for quite early and quite recent dates, say, 1935 and 1995. Any observed narrowing or widening of rivers and estuaries over such a 60-year interval would provide some indication of the reliability of the prediction of sinking there.

Another reading fragment that has been cited[10] as a failed Earth-change prophecy is this one:

Q Will the Earth upheavals during 1936 affect San Francisco as it did in 1906?

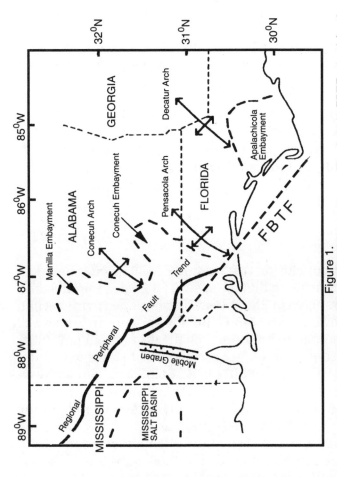

Figure 1.

Structural features in the subsurface of the southwestern Alabama area. The letters FBTF stand for the interpreted location of the Florida-Bahamas Transfer Fault. (Redrawn and simplified from G. MacRae and J.S. Watkins, 1995, *Journal of Geophysical Research*, v. 100, no. B9, Figs. 3 and 4; copyright by the American Geophysical Union.)

A This'll be a baby beside what it'll be in '36!
270-30
February 13, 1933

The person who asked this question must have been
aware of reading 5748-6 below that had been given seven
months earlier:

Q What will be the type and extent of the up-
heaval in '36?
A The wars, the upheavals in the interior of the
Earth, and the shifting of same by the differentia-
tion in the axis as respecting the positions from the
Polaris center.
5748-6
July 1, 1932

Any conclusion that reading 270-30 failed is based on
the supposition that the answer addresses the effects on
San Francisco of the upheavals mentioned in the second
reading quoted (5748-6). But the answer can just as well
be assumed to be addressing the disparity in power between one near-surface earthquake in 1906 and massive
upheavals deep within the Earth in 1936. The "upheavals in the interior of the Earth" mentioned in reading
5748-6 would have been many orders of magnitude
more powerful than was the shaking, related to sudden
slippage along the San Andreas fault, that led to the destruction of San Francisco in 1906. Supporting this interpretation is the answer of the following 1934 reading,
in which the psychic source refuses to consider seismotectonic effects on the Pacific coast, due to upheavals in
the interior of the Earth in 1936.

Q Are details of the Earth's eruptions in 1936 so

fixed that you can give me an outline of the Pacific Coast area to be affected, along with precautionary measures to be exercised during and after this catastrophe?

A All of these are, as is ever on or in such an activity, dependent upon individuals or groups who are in or keep an attitude respecting the needs, the desires, the necessary requirements in such a field of activity. That some are *due* and *will* occur is *written*, as it were, but—as we find—as to specific date or time in the present this may not be given.

270-32
June 12, 1934

This is because a time lag of unspecified length is required between the time of the upheavals in the interior of the Earth (1936) and their robust expression at the Earth's surface (beginning presumably in 1998, as explained later on).

A final "failed prediction" is the following:

And Poseidia will be among the first portions of Atlantis to rise again. Expect it in sixty-eight and sixty-nine ('68 and '69); not so far away!

958-3
June 28, 1940

The hypothetical, submerged, subcontinental-sized crustal fragment, called Poseidia in the readings, did not rise above the ocean surface in 1968 or 1969. But perhaps reading 958-3 meant that the part of Atlantis called Poseidia would *begin* to rise in 1968 or 1969. This interpretation fits in with the mention of Atlantis rising, as given in the reading below:

Before that we find the entity was in the Atlantean land, when there were the constructive forces as to the activities of the children of the Law of One—in all of those influences during the periods when the land was being broken up. We find the entity was as the leading influence for the considering of ways and means in which there would be the preserving of records, as well as ways, means and manners in which either the few or the numbers might be preserved from the destruction of the lands. It would be well if this entity were to seek either of the three phases of the ways and means in which those records of the activities of individuals were preserved—the one in the Atlantean land, that sank, which will rise *and is rising again* [emphasis added]; another in the place of the records that leadeth from the Sphinx to the hall of records, in the Egyptian land; and another in the Aryan or Yucatan land, where the temple there is overshadowing same.

2012-1
September 25, 1939

Thus, our interpretation of these two readings is that the part of Atlantis that contains certain records was rising in 1939 and, as of 1968 or 1969, Poseidia was also beginning to rise.

Another factor that enters into judging the accuracy of psychic information is that sometimes the event is foreseen but the timing remains uncertain. Some psychics will admit that they sometimes visualize coming events but have a hard time measuring when they will occur, as we define time. When Cayce was asked when we can expect the Earth changes, he sometimes said, in effect, only God knows, as in this reading:

A As to times and places and seasons, as it has indeed been indicated in the greater relationships that have been established by the prophets and sages of old—and especially as given by Him, "As to the day and the hour, who knoweth? *No one, save the Creative Forces.*"

416-7
October 7, 1935

These changes in the Earth will come to pass, for the time and times and half times are at an end[11] and there begin these periods for the readjustments. For how hath He given? "The righteous shall inherit the Earth." Hast thou, my bretheren, a heritage in the Earth?

294-185
June 30, 1936

Finally is the question of just who is providing the information we find in the Cayce readings. No less than six sources—as explained by the readings themselves—are described in the Appendix. Only rarely did Cayce act as a medium—one who gives voice to information from a disincarnate spirit—as did the famous Irish psychic Eileen Garrett (whose source was identified as Uvani). Nonetheless, there were a few readings in which discarnate spirits spoke through Cayce. One of them identified himself as the archangel Michael. On such occasions Cayce's voice is said to have changed dramatically and virtually blasted his listeners.

A second entity who came through was identified in one of the most significant Earth-change readings. We know this because halfway through the reading we find the statement, "I, Halaliel, have spoken." But who is Halaliel? When Cayce was asked this question, the response was:

A One in and with whose courts Ariel fought when there was the rebellion in heaven. Now, where is heaven? Where is Ariel, and who was he? A companion of Lucifer or Satan, and one that made for the disputing of the influences in the experiences of Adam in the Garden.

262-57
January 7, 1934

Several other readings shed further light on the Halailel question:

Q To what extent are the Masters of the Great White Brotherhood directing the activities of Edgar Cayce? Who are the Masters directly in charge?
A Messengers from the higher forces that may manifest from the Throne of grace itself.
Q Who are the Masters directly in charge? Is Saint Germain—
A—[interrupting] Those that are directed by the Lord of lords, the King of kings, Him that came that ye might be one with the Father.
Q Is Saint Germain among them? Who is Halaliel?
A These are all messengers of the Most High. Halaliel is the one who from the beginning has been a leader of the heavenly host, who has defied Ariel, who has made the ways that have been heavy—but as the means for the understanding.

254-83
February 14, 1935

Q How high is this source that this information is being given from?

A From the universal forces, and as emanated through the teacher that gives same—as one that has been given—Halaliel.

443-3
January 8, 1934

Q If Edgar Cayce has ever had controls, does he know who they are?

A Anyone may speak who may seek, if the entity or the soul's activities will allow same; or if the desire of the individuals seeking so overcommands as to make a set channel.

Q What entity is giving this information now?

A Being directed, as has been indicated, from the records through Halaliel.

507-1
February 3, 1934

These readings suggest that reading 3976-15, communicated by Halaliel through the entranced Cayce, contained information from the highest of forces, the "universal forces." They also claim that the entity Halaliel is indeed a messenger, even a leader of the heavenly host, which would account for his being a source[8] for Earth-change reading 3976-15. Below are the remaining two Earth-change paragraphs of Halaliel's discourse. They precede the extract quoted above:

As to the material changes that are to be as an omen, as a sign to those that this is shortly to come to pass—as has been given of old, the sun will be darkened and the Earth shall be broken up in divers places—and *then* shall be *proclaimed*— through the spiritual interception in the hearts and minds and souls of those that have sought His

way—that His star has appeared, and will point the
way for those that enter into the holy of holies in
themselves. For, God the Father, God the Teacher,
God the Director, in the minds and hearts of men,
must ever be in those that come to know Him as
first and foremost in the seeking of those souls ...

As to the changes physical again: The Earth will
be broken up in the western portion of America.
The greater portion of Japan must go into the sea.
The upper portion of Europe will be changed as in
the twinkling of an eye. Land will appear off the
east coast of America. There will be the upheavals
in the Arctic and in the Antarctic that will make for
the eruption of volcanoes in the Torrid areas, and
there will be the shifting then of the poles—so that
where there has been those of a frigid or the semi-
tropical will become the more tropical, and moss
and fern will grow. And these will begin in those
periods in '58 to '98, when these will be proclaimed
as the periods when His light will be seen again in
the clouds. As to times, as to seasons, as to places,
alone is it given to those who have named the
name—and who bear the mark of those of His calling and His election in their bodies. To them it shall
be given.

                                        3976-15
                                January 19, 1934

It is impossible to tell whether any subsequent Earth-
change readings were given with the help of Halaliel.

But what about the accuracy of that Halaliel reading?
Cayce was asked what changes could be expected that
year, 1934. The reply was startling:

A The Earth will be broken up in many places.

The early portion will see a change in the physical aspect of the west coast of America. There will be open waters appear in the northern portions of Greenland. There will be new lands seen off the Caribbean Sea, and dry land will appear ... South America shall be shaken from the uppermost portion to the end, and in ... Tierra del Fuego *land,* and a strait with rushing waters.

3976-15
January 19, 1934

Since the question specified 1934, either the predictions were wrong or Halaliel simply ignored the "this year" specification. I believe that the latter is the correct interpretation because an earlier part of the same reading clearly states that "these [physical changes] will begin in those periods in '58 to '98, when these will be proclaimed as the periods when His light will be seen again in the clouds." Thus, the context of the entire reading suggests that the "this year" qualifier in the above question translates to "this period," as in " '58 to '98." That is the time period during which all of the Earth changes foreseen in this reading would *begin* to occur. Halaliel neither said nor implied that all of the worldwide Earth changes he described would occur in just the one year, 1934. At the same time, however, it is certainly true that "the early portion" of the 1958 to 1998 period did *not* see "a change in the physical aspect of the west coast of America," at least in the sense of Earth changes. But in terms of "the world changes," *whenever* they start to be clearly visible at the surface, we have reading 1152-11 that says that west coast cities like Los Angeles will be destroyed before New York City, thus agreeing with the timing of 3976-15 in which the early portion of the changes will see a change in the physical aspect of the west coast.

Reading 3976-15 accounts for eight of the 24 Earth-change predictions discussed in this book. We will examine all of these dire forecasts in greater detail and see if there are correlations between them and the results of recent, geophysical studies. We'll begin with the event that would provide the driving force for most all of the other Earth changes visualized by Cayce—a shift in the poles of the Earth.

2

Pole Shift

Shifts in the Earth's poles, with attendant rising and sinking of large-sized blocks of the crust, is a recurrent theme in the Cayce readings. Here are the readings that predict a pole shift and describe the nature of some of the Earth-change events that will be associated with it. (Words italicized in each reading are for phrase-finding purposes only.)

Q What great change or the beginning of what change, if any, is to take place in the Earth in the year 2000 to 2001 A.D.?

A When there is a *shifting of the poles.* **Or a new cycle begins.**

826-8
August 11, 1936

As to the changes physical again: The Earth will be broken up in the western portion of America. The greater portion of Japan must go into the sea. The upper portion of Europe will be changed as in the twinkling of an eye. Land will appear off the east coast of America. There will be the upheavals in the Arctic and in the Antarctic that will make for the eruption of volcanoes in the Torrid areas, and there will be *shifting then of the poles*—so that where there has been those of a frigid or the semi-tropical will become the more tropical, and moss and fern will grow. And these will begin in those periods in '58 to '98, when these will be proclaimed as the periods when His light will be seen again in the clouds. As to times, as to seasons, as to places, ALONE is it given to those who have named the name—and who bear the mark of those of His calling and His election in their bodies. To them it shall be given.

3976-15
January 19, 1934

In the record chambers [in Egypt] there were more ceremonies than in calling the peoples at the finishing of that called the pyramid. For, here those that were trained in the Temple of Sacrifice as well as in the Temple Beautiful were about the sealing of the record chambers. For, these were to be kept as had been given by the priests in Atlantis or Poseidia . . . when the records of the race, of the developments, of the laws pertaining to One were put in their chambers and to be opened only when there was the returning of those into materiality, or to Earth's experience, when the change was imminent in the Earth; which change, we see, begins in '58 and ends with the changes wrought in the upheavals and *the shifting of the poles,* as begins then the reign in '98 (as time is counted in the present) of those influences that have been given by many in the records that have been kept by those sojourners in this land of the Semitic peoples.

378-16
October 29, 1933

Q Three hundred years ago Jacob Boehme decreed Atlantis would rise again at this crisis time when we cross from this Piscean Era into the Aquarian. Is Atlantis rising now? Will it cause a sudden convolution and about what year?

A *In 1998 we may find a great deal of the activities as have been wrought by the gradual changes that are coming about.* [Emphasis added.] These are at the periods when the cycle of the solar activity, or the years as related to the sun's passage through the various spheres of activity become paramount ... to the change between the Piscean and the Aquarian age. *This is a gradual, not a cataclysmic activity in the experience of the Earth in this period.* [Emphasis added.]

1602-3
September 22, 1939

My interpretation of these readings is that the 1958-to-1998 period of gradually increasing Earth changes will end with upheavals in the Arctic and Antarctic. Leading up to, or coinciding with, these upheavals we may expect that western North America will be broken up, that much of Japan will be submerged, that land will appear off the east coast of America, that northern Europe will be changed "in the twinkling of an eye," and that nu-

merous volcanoes will erupt between the tropics of
Cancer and Capricorn (the Torrid Zone). These Earth
changes will culminate in a modest pole shift. Although
not deemed cataclysmic in reading 1602-3, the predicted
events are unprecedented by historical standards. However, if gradual enough, they could be less catastrophic
than other past global events inferred from the geological record, such as the vast outpourings of lava in Sibe-
ria, India, and the Pacific Northwest of the United States
in eons gone by.

What would be the underlying cause of such a pole
shift? A reading given in 1932 ascribes the primary cause
to a sudden instability in the rotating Earth. It said that
the pole shift would be related to:

> . . . the catastrophes of outside forces to the
> Earth in '36, which will come from the shifting of
> the equilibrium of the Earth itself in space, with those
> of the consequential effects upon the various por-
> tions of the country—or world—affected by same.
>
> 3976-10
> February 8, 1932

It is difficult to tell from the readings if this instability
is due only to outside forces or to upheavals in the interior of the Earth, as mentioned in the next reading:

> Q What will be the type and extent of the up-
> heaval in '36?
> A The wars, the upheavals in the interior of the
> Earth, and the shifting of same by the differentia-
> tion in the axis as respecting the positions from the
> Polaris center.
>
> 5748-6
> July 1, 1932

Cayce's reference to "upheavals in the interior of the Earth" was given decades in advance of contemporary scientific notions about plumes of hot material ascending through the mantle from deep within the Earth. Only recently, for example, did three geophysicists[1] publish their laboratory research on cycles of activity within the Earth's mantle and core. They proposed that the cycles are due to variations in the thickness of the warm layer at the base of the mantle, as the layer alternates between two kinds of activity. In the quiescent phase, little flow occurs in the warm layer over long periods of time. During this phase, the layer becomes thicker and thicker as heat from the Earth's core diffuses outward. The active phase starts when the thickening layer becomes dynamically unstable. Ultimately, hot material erupts from the layer and moves toward the surface of the Earth where it causes widespread crustal deformations and volcanic eruptions. These scientists believe that these surface effects should not necessarily occur at exactly the same time. All of the surface effects will vary according to the strength of the plumes of hot material arising from the lower mantle.

A final point is made that a plume event could have surface effects ranging from normal hot-spot volcanism to an extremely sudden and explosive upheaval. These scientists say that it is possible that the Earth could experience several such events at the same time. By inference from their laboratory experiments, the lower mantle could become unstable at a number of locations at the same time. Each of the surface upheavals might have a different ascent history and might be preceded by an intensification of hot-spot volcanism in other parts of the world. (This last hypothesis resonates with the part of reading 270-35 that links predicted significant eruptions of Mt. Vesuvius, Italy, or Mt. Pelée, Martinique, with earthquakes in California and Nevada up to three months later.)

But what evidence is there, beyond the laboratory, that one or more plumes might be on the way from Earth's core-mantle boundary to the surface? In 1993, M.E. Wysession summarized[2] recent research on this subject as follows:

> **A new picture of the dynamics at the base of the Earth's mantle is emerging from the integrated work of seismologists, geodynamicists, and geomagnetists. The structure at the core-mantle boundary (CMB) appears to be influencing convective motions both in the solid silicate mantle and in the liquid-iron outer core. Such processes are apparently at work at the CBM beneath eastern Indonesia, where very slow seismic velocities suggest an unusually hot region and could signal the birth of a mantle plume. In addition, geomagnetic modeling indicates that rising plumes could exist here in the core ...**

See Fig. 2 for a pictorial summary[3] of current thoughts about the sources and dynamics of thermal plumes. A new geophysical methodology—teleseismic tomography—is now being used to locate mantle plumes. An ancient mantle plume that ascended beneath the French Massif Central has just been unequivocally documented by M. Granet, et al., in *Earth and Planetary Science Letters* (1995, v. 136, pp. 281-296).

Also, *Science* magazine for September 14, 1984, reported that "something seems to have happened within the Earth's core that jerked the magnetic field in 1969 during its slow drift across the globe." Whether any of the hypothetical motions of the Earth's core that jerked

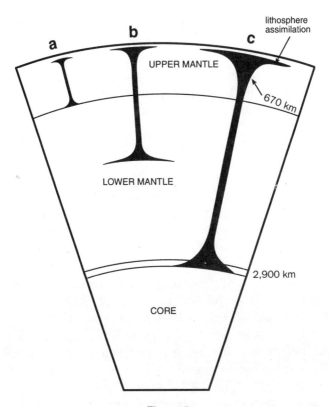

Figure 2

Earth's mantle rocks lie between thin lithospheric plates at the surface and the molten iron core 2,900 kilometers (1,802 miles) below. They are under such high pressures and temperatures that they behave more like molasses than solid rocks. Geophysicists currently think that the lower mantle—below the 670-kilometer (416-mile) boundary where increasing pressure transforms the crystal structure of the mantle rock—is much stiffer than the upper mantle. In this schematic idealization, rising plumes of mantle rock can originate (a) at the thermal boundary layer between the upper and lower mantle, or (c) at the core-mantle boundary. Some plumes may originate (b) in zones of thermal instability within the lower mantle as well. Powerful plume (c) has pushed so close to the surface that it is interacting with the lithosphere, assimilating crustal material. (Adapted from Figure 1 of DePaolo, and others, 1991, "Physics and Chemistry of Mantle Plumes," *EOS,* May 21; copyright by the American Geophysical Union.)

the magnetic field could have reflected somehow the "upheavals in the interior of the Earth" described in the reading is anybody's guess. We note in passing, however, that the beginning of submerged Poseidia's rise from the ocean was to be expected in "sixty-eight or sixty-nine" (958-3, June 28, 1940) and that the sudden jerk in the magnetic field in 1969 might just have reflected some aspect of a core and/or mantle-plume upheaval. Since 1969, geomagnetic jerks occurred again in 1978 and 1991, and "they are believed to be internal in origin and have implications for the understanding of the fluid motions in the core which create them, or for lower-mantle conductivity, or both" (*Earth and Planetary Science Letters*, 1996, v. 137, p. 189.)

We next ask what information there might be for a (slight) shifting of the poles of the Earth in 1936 "by the differentiation in the axis as respecting the positions from the Polaris center." W. Markowitz[4] examined the mean latitudes of the five International Latitude Service (ILS) stations for the mean "epochs" of 1903, 1909, 1915, 1927, 1932, 1938, and 1952. These stations use astronomical observations to infer periodic oscillations of the instantaneous pole about its mean position, and to infer other polar motions as well. Accounting for the unequal time differences between observational epochs, the greatest differences in latitude and longitude positions for all five ILS stations were found to occur between mean epochs 1932 and 1938. Perhaps these differences reflect a *shifting of the equilibrium of the Earth itself in space* in 1936.

A recently announced discovery allows us to consider a completely different aspect of the statement in reading 3976-10 regarding shifting of the equilibrium of the Earth in space in 1936. In a front-page article in the *New*

York Times of July 18, 1996, William Broad reported that three scientific groups had found strong evidence that Earth's inner core of solid iron is spinning freely within the molten outer core, making the inner core "virtually a planet within a planet." The evidence comes from study of seismic waves from earthquakes and explosions that travel through Earth's inner core. Of great interest to us is Broad's statement that, "For unknown reasons, the axis of the core . . . is tilted a few degrees with respect to Earth's north-south axis of rotation . . . "

Let's carefully re-read the answer of reading 5748-6, quoted immediately above. In response to a question about the type of upheaval to be expected in 1936, the reading said:

> **The wars, the upheavals in the interior of the Earth, and the shifting of same by the differentiation in the axis as respecting the positions from the Polaris center.**

Now if "of same" refers to the interior of the Earth, the reading is saying that the rotational axis of the iron core (the "interior of the Earth") became "differentiated" from its usual position of alignment with Earth's rotational axis. Researchers now think that the rotational axis of Earth's solid inner core is tilted at about 10 degrees to that of Earth's. If that degree of tilt suddenly occurred, or began to occur, in 1936, it could certainly engender rising thermal plumes and profound geophysical effects at Earth's surface over time. If so, the time lag between axial differentiation and the *beginning* of surficial effects, according to the readings, would amount to between 22 and 62 years, or from 1958 to 1998. And shifting of Earth's rotational (?) axis in 1998 would be a follow-on to the catastrophic shift of the equilibrium of the [inner] Earth itself in space in 1936 (3976-10).

Putting the above laboratory, theoretical, and observational information together, one can say that there may be at least some scientific basis for the psychically derived statements about upheavals in the interior of the Earth starting in 1936 and leading to rising thermal plumes that will culminate in the "breaking up of some conditions in the South Sea . . . that's South Pacific" (311-8, 1932), in eruption of volcanoes in the Torrid Zone, and in "upheavals in the Arctic and in the Antarctic." According to the readings, these activities will coincide with a shifting in the poles in 1998 to 2001 A.D. But what will be the nature of this shifting of the poles? Will Earth experience true polar wander,[5] consisting of a shifting of the entire mantle relative to Earth's spin axis, or will it experience displacement of the spin axis itself? While not answering this question with certainty, one reading is instructive about a previous pole shift:

> You see, with the changes—when there came the uprisings in the Atlantean land . . . with *the turning of the axis* [emphasis added], the white and yellow races came more into that portion of Egypt, India, Persia and Arabia.
>
> 364-13
> November 17, 1932

Here, it would definitely seem that the psychic source identified spin-axis displacement as the type of polar shift. But we must wonder about the conceptualization in the reading that led to Cayce's words "with the turning of the axis." He may have meant exactly *that*, that Earth's spin axis moved with respect to the stars. But if Earth's spin axis is fixed with respect to the stars, as most scientist believe, then it is actually the Earth that slowly tumbles like a rolling ball beneath the pole, where the spin axis meets the surface, to create "polar shift."

By reference to "the uprisings in the Atlantean land," the pole shift mentioned above probably began around 52,718 B.P. and continued during the period when men began to use "*explosives* that might be carried about . . . [and] . . . with these destructive forces, we find the first turning of the altar fires into that of sacrifice of those that were taken in the various ways, and human sacrifice began" (364-4). These activities "brought about the first upheavals" (1292-1). The reading below seems to describe the same pole shift as given in 364-13 above, but it may describe a later one, depending upon which "uprisings in the Atlantean land" reading 364-13 is talking about.

> **The entity then was among those who were of that group who gathered to rid the Earth of the enormous animals which overran the Earth, but ice, the entity found, nature, God, changed the poles and the animals were destroyed . . .**
>
> **5249-1**
> **June 12, 1944**

According to readings 262-39 and 364-4, this gathering occurred in 50,722 B.C. (52,718 B.P.). But this date would most probably mark only the start of ice formation in response to pole shift because it takes millennia to build an ice sheet. For example, the maximum extent of North American ice-sheet occurred about 20,000 years ago. And if ice did in fact "change the poles," it probably did so at the time of greatest ice thickness. For reasons given later, we will postulate that an *ice*-induced pole shift occurred about 19,400 B.P., in conjunction with a man-made destruction that split Atlantis into five islands (877-26) and sent part of that continent into the

Sargasso Sea (364-11). Key to this line of thought is a requirement that the North Pole—*before the shift at 19,400 B.P.*—was in northern Greenland. Information suggesting that this could have been so comes from a leading numerical modeler of the growth and movement of the most recent northern hemisphere ice sheets. The modeler has found that in order for his model to fit the known ice-sheet margins around the time of the glacial maximum at 20,000 B.P., he must use a *climate* center in northern Greenland, not the one centered at the present North Pole. This is the justification for our assumption that the geographic pole was located in northern Greenland before the inferred pole shift of 19,400 B.P. Note also that for reading 5249-1 to make sense, only by way of reincarnation would "the entity" have come to realize—over many lifetimes—that "ice . . . nature, God, changed the poles and the animals were destroyed." And it may have been a pole shift—from a point elsewhere on Earth to one in northern Greenland—at the time of this gathering that got the Wisconsin Age ice sheets building.

As for those "enormous animals" that concerned the group in reading 5249-1, we note that North America and Europe were inhabited by great game animals fully as varied and even more impressive than those of modern East Africa. A partial list includes elephants, two of the four species of which exceeded modern elephants in size. The tall imperial mammoth of the Great Plains stood 14 feet high at the shoulders. Throughout the forests, mastodons browsed in great herds. Horses abounded, one as large as a modern draught horse. There were seven species of buffalo, and one was an enormous beast with a horn spread of 14 feet. The musk ox, woolly rhinoceros, great ground sloth, bear, wild pig, camel, and giant beaver also abounded. Accelerated extinction of most of the giant mammals began toward the end of the Wisconsin ice age, about 25,000 B.P., and culminated 12,000 to 10,000 B.P., as climatic and environmental changes delivered a fatal blow to many large vertebrate species.[6] In their study, "Ice Cores and Mammoth Extinction,"[7] A.M. Lister and A.V. Sher state that around 11,500 B.P. "there was a second major [climatic/ecological] event evidenced by the appearance of a myriad of thaw lakes all over the unglaciated Arctic, marking the final demise of the tundra-steppe biome." *This* was the moment of greatest loss (by extinction) of the hordes of great beasts that had been roaming parts of the Earth. And if the readings are correct, it also marks the time, around 11,900 B.P., when the last of Atlantis went down in a spasm of volcanism.

Getting back to a pole shift in 19,400 B.P., we note that this was a time of sudden (geologically speaking) imposition of a different density structure in the outer shell of the northern hemisphere due to (1) the large masses of ice that had built up on the continents, (2) the large masses of water that had been removed from the ocean basins to build those ice sheets, and (3) the mass of the Atlantean subcontinental fragment that sank into the area of the Sargasso Sea at this time. The mechanism for pole shift from such conditions is similar to an explanation[8] of how the Earth's axis of rotation can be shifted by the sinking of cold, dense slabs of mantle in subduction zones such as those surrounding the Pacific Ocean. The Earth's axis of rotation may shift as the equator moves closer to the denser, sunken subcontinental fragment, providing the most stable arrangement for a rotating Earth. For the case of our hypothesized pole shift at 19,400 B.P., we have additional contributing conditions, but the principle is the same.

Thus, we have the sequence: (1) shift of the North Pole to northern Greenland, from an unknown position on

Earth, beginning about 52,718 B.P., (2) continued development of northern hemisphere ice sheets until 19,400 B.P., (3) shift of the North Pole from northern Greenland to its present position, beginning about 19,400 B.P., and (4) accelerating melting of the ice sheets and deaths of large animals until the final demise of the animals about 12,000 to 10,000 B.P.

Assume that the buildup of the Wisconsin Age ice sheet did in fact "change the poles" by displacing the Earth's axis of rotation. From 52,718 B.P. (Cayce's 50,722 B.C.; see large dot in Fig. 3) to 30,000 B.P., the global climate experienced ups and downs in a relatively warm stage (Fig. 3). But after 30,000 B.P. the climate became cold enough to produce the expanded ice sheets of the most recent continental glaciation. In Illinois, around 48,000 B.P., for example, what geologists call the Sangamon soil had been developing, but over the following millennia, it began to be covered with windblown proglacial silts originating from summer melt waters from the expanding North American ice sheet.

Around 25,000 B.P. glacial ice entered northeastern Illinois, and by 20,000 B.P. it had reached its maximum extension. Extensive melting of the ice sheet began to be apparent around 18,500 B.P. in the midwest. The area now occupied by Chicago was deglaciated by 13,500 B.P., about the time that Earth's climate entered what geologists have designated as the present interglacial, and the Straits of Mackinac, Michigan, became ice free around 11,000 B.P. A roughly similar chronology holds for the expansion and contraction of the European ice sheet of this period.

Ice cores show that the Greenland ice sheet existed for at least 110,000 years or so, and the Antarctic ice sheet for at least the last 150,000 years. Huge volumes of ice were added upon them and beyond their margins during colder climatic phases (B, D, and F of Fig. 3). For ex-

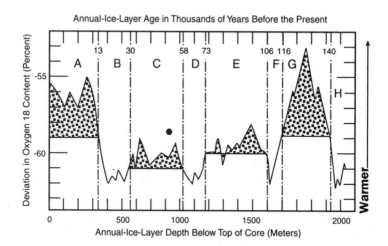

Figure 3

A "Rosetta-stone" for deciphering Earth's climate over the last 150,000 years. The graph shows the change in oxygen-18 isotope content of annual ice layers versus depth in a 2,000-meter-long core from the Antarctic ice sheet (Vostok research station [78°28'S; 106°48'E].) By counting annual ice layers and measuring their oxygen-18 content, an indicator of atmospheric temperature, scientists were able to construct a continuous record of the warmth of the global atmosphere. The isotopically richest layers (lying around -55 through -57 percent) represent the warmest ancient global atmospheres (panels G and A on the graph, respectively). Panel A correlates with the present Holocene epoch that began roughly 10,000 years ago, and panel G with the peak warmth of the last interglacial phase of the Pleistocene epoch. Panel H represents the last part of the previous glacial phase. Panels E and C correlate with relatively warm interstadials, or climatic episodes within a glaciation during which a secondary recession or still-stand of continental glaciers takes place. (Adapted from C. Lorius, et al., 1985, *Nature,* 316: pp. 591-596). The large dot within interstadial C denotes the approximate time of the meeting of the group that "gathered to rid the Earth of the enormous animals . . . "

ample, at the height of the last glaciation, around 20,000
B.P., a lobe of the North American ice sheet reached central Indiana, covering the now Indianapolis area with
about 1,700 feet of ice. And this was at a point only 28
miles from the edge of the ice sheet.[9] Recent computer
models of the buildup and movement of the last north-
ern-hemisphere ice sheet require its thickness at the
North American spreading center (Hudson Bay?) to have
been at least 12,500 feet.

It is clear that the pole shift, inferred from the readings to have occurred around 19,400 B.P., was not sufficient to destroy Earth's two existing ice sheets because
continuous annual layers of ice have been cored from
both ice masses back beyond 100,000 B.P. This suggests
limits on the extent of a future pole shift. To expand on
this subject, consider a modest pole shift in which the
North Pole moves along the 150°W meridian to a point
only 10° south of its present position. Such a move would
result in open waters appearing "in the northern portions of Greenland" (3976-15) because that area would
be roughly 8° farther from the pole than now, permitting
pack ice to melt there. Northernmost Norway would experience, *climatically,* the latitude of present-day
Leningrad. That is, present climatic zones would be displaced *northward* along the 30°E meridian by 10° of latitude. Such displacement would find the present tropical
rain forest and savanna (subtropical rainy climate) belts
of middle Africa occupying most of the present subtropical desert belt of northern Africa and the Arabian Peninsula. Although we're postulating only a 10° pole shift
here, it would seem to agree with the botanical features
mentioned in reading 3976-15: "*. . . there will be the shifting then of the poles—so that where there has been those
of a frigid or the semitropical ['subtropical' in current us-
age] will become the more tropical, and moss and fern*

will grow." Although mosses and ferns can grow along-side one another, here we would expect moss to re-present harsh, frigid growing conditions and fern to represent tropical, rain-forest conditions. Thus, in our 10° pole-shift scenario, an Arctic island like Spitsbergen would suddenly experience enough of a warmer climate to cause its glaciers to melt and Arctic tundra moss to grow on the exposed surfaces, while the sudden imposition of a rain-forest climate on large portions of the deserts of North Africa and the Arabian Peninsula would permit ferns to grow there. Finally, note that the South Pole would move to the middle of Queen Maud Land and most all of the Antarctic ice sheet would be preserved. In Greenland, only the southeasternmost portions of the ice sheet might melt.

The foregoing 10° shift is but one example of a large number of modest pole shifts that could be considered, any one of which would preserve most of each of the two ice sheets, as the ice-core records indicate happened in spite of the past pole shift mentioned in the readings. And could one expect "eruption of volcanoes in the Tor-rid Zones" (3976-15) to be triggered by such modest pole shifts? Yes again, as explained later on.

Now how, once more, do we get from ice sheets to po-lar shift, as opposed to invoking rising mantle plumes to produce pole shift? The huge mass of an ice sheet forces downward the Earth's underlying crust and mantle. This slab of depressed rock, together with the ice mass itself, results in a broad area of the Earth's outer shell that has a higher density than existed prior to ice buildup. Such a positive density change on land will be enhanced by a *negative* density change over the ocean basins due to unloading of the crust by the transfer of ocean-water mass to the growing ice sheets. As the equator moves closer to the area(s) of higher relative density (see Fig. 4),

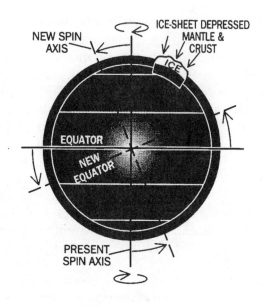

Figure 4
The Earth's axis of rotation can shift in response to movement of the
equator toward an area of higher density, suddenly imposed by ice-
sheet buildup and related sinking of the underlying crust and mantle.

the Earth's axis of rotation may shift. Thus, climate in
part ("ice . . . nature" in reading 5249-1) is seen here as
having contributed to the pole shift around 19,400 B.P. A
significant role for rapid climate change, as occurs in an
ice age, has recently been suggested by Rampino, et al.:

> . . . variations in climate lead to stress changes
> on the Earth's crust—for instance, by loading and
> unloading of ice and water masses and by axial
> and spin-rate changes that might augment volca-
> nic (and seismic) potential.[10]

Because today's climate is no longer glacial but is instead approaching the warmth of some previous interglacial climates, one may conjecture that continued deglaciation will lead again to polar shift, but a shift caused this time by a mechanism that is the mirror image of that which occurs in response to ice-sheet formation. Reading 3976-15 addressed this potentiality as one of many widespread geophysical events that will begin to occur in the period 1958 to 1998: *"There will be the upheavals in the Arctic and in the Antarctic that will make for the eruption of volcanoes in the Torrid areas, and there will be shifting then of the poles—so that where there has been those of a frigid or the semitropical will become more tropical . . . "* As the global climate warms and the Greenland and Antarctic ice sheets melt at increasing rates, the ice loads will diminish and the crust and underlying mantle will rebound (producing "upheavals"). A recent model of this process[11] predicts that vertical uplift presently underway over a vast portion of West Antarctica is in excess of 20 mm/yr. This is a huge shift in mass and density, and modeling suggests it is producing "quite substantial" changes in the presently very minor drift of the Earth's spin axis (the so-called "secular" polar motion). Geoscientists are calling for a precision surface-elevation measurement campaign around the ice sheets to help them refine their models of the mantle rebound now underway.

Potential precursors of "upheavals" in Antarctica may be the currently active volcanoes found in both the Transantarctic Mountains and Marie Byrd Land. Active volcanism has also been documented *beneath* the West Antarctic ice sheet, landward of the Ross Ice Shelf in a "region thought to contain the inland edge of hot, actively extending lithosphere"[12]

As the Earth's equator attempts to respond to the

changing density imperatives due to (1) mantle plumes rising to the surface, (2) crustal upheavals in the Arctic and Antarctic areas, and (3) crustal uplift in the areas still undergoing rebound from the last continental glaciation, the outer shell of the Earth can be stressed to failure by migration of the equatorial bulge in the mantle. It is the viscous mantle that deforms to reach a new configuration of the Earth's figure, which is an oblate ellipsoid of revolution. Under this scenario for movement of the Earth's spin axis, accelerated eruption of volcanoes in the Torrid Zones can be expected due to migration of the equatorial mantle bulge. Thus, reading 3976-15 may be interpreted to suggest that impending polar shift is presaged by events such as the 1991 Mt. Pinatubo eruption in the Philippines and by strong, deep quakes associated with subducting slabs of oceanic mantle in the torrid zone, lying between the tropics of Cancer and Capricorn. A magnitude 8.2 deep quake occurred near Guam (Mariana Trench) in August of 1993, and the largest deep quake ever recorded (8.3 magnitude) struck beneath Bolivia in June 1994. Both quakes resulted from subducting slabs located in the Torrid Zone.

Now consider again the scenario of pole shift due to outside forces affecting the equilibrium of Earth in space and causing rising mantle plumes ("upheavals in the interior of the Earth"). These rising plumes could lead to a shifting of the entire mantle relative to the Earth's spin axis. In either the melting-ice-sheet-and-migrating-mantle-bulge scenario or the worldwide-mantle-shift scenario, widespread crustal changes would result. (As reading 3976-15 states, "the Earth shall be broken up in divers places.")

Eruptions of volcanic calderas might be looked to as sensitive indicators of the beginning of significant mantle movements. Caldera eruptions are among the largest class of volcanic explosions, ejecting tens to hundreds of cubic kilometers of magma. When such large volumes are ejected from volcanoes, the ground collapses into the empty spaces to form huge depressions, called calderas.

Recently, three well-known calderas have simultaneously begun to act up.[13] Campi Flegrei, near Mt. Vesuvius, Italy, is experiencing uplift of its floor. Rabaul, a torrid-zone caldera in New Guinea, produced twin volcano eruptions in September 1994. Long Valley, in eastern California, has been displaying long-period quakes and a rising dome of hot magma. These same calderas were also particularly restless in the early 1980s. Wyoming's Yellowstone caldera also shows signs of unrest, the most evident being historic uplift and subsidence of the ground surface.[14] Is the mantle's pulse accelerating? Rabaul had been restless for nine years before erupting in 1994. The last eruption of Campi Flegrei, however, was in 1538. Long Valley, although restless, has never had an historical eruption as large as Rabaul's.

If in fact it is not delusional for us to keep an eye on torrid-zone volcanoes for a pole-shift early-warning system, we have a couple of awesome examples for baseline evaluations of future caldera eruptions. The two largest caldera-type eruptions in the historical record were in Indonesia. Krakatoa, which started off with a good-sized eruption in May of 1883, quieted down, and then blew itself away in August. Tambora, however, took three years to build to its 1815 eruption, the largest ever recorded. The final years of the 1958-to-1998 period may provide us with enough climatologic and geodynamic information to allow a more conclusive evaluation of the potential for the shifting of the poles.

Finally, when considering the relative importance of crustal rebound by melting of the ice sheets to the effects of mantle plumes in shaping the upcoming pole shift, it would seem that the effects of mantle plumes, and perhaps other types of "upheavals in the interior of the Earth," would be, far and away, the most important. This could mean too—for the surface of the Earth—episodes of volcanism, ash falls, ash-darkened skys, earthquakes, and crustal movements unlike any experienced by those now living.

As we approach 1998, when Earth changes are to begin in earnest, we are intrigued by the following reading, given for a woman who was one of "the eight souls in the ark—as the wife of a son of Noah."

> **For as has been given from the beginning, the deluge was not a myth (as many would have you believe) but a period when man had so belittled himself with the cares of the world . . . as to require that there be a return to his dependence wholly— physically and mentally—upon the Creative Forces.**
>
> **Will this entity see such again occur in the Earth? Will it be among those who may be given those directions as to how, where, the elect may be preserved for the replenishing again of the Earth?**
>
> **Remember, not by water—for it is the mother of life in the Earth—but rather by the elements, fire.**
>
> **3653-1**
> **January 7, 1944**

Will the elemental fire mentioned express itself in the form of volcanic upheavals? Other readings suggest that as we enter the post 1958-1998 period, upheavals in the polar regions will cause volcanoes to erupt in the Torrid Zones and that the sun will be darkened—presumably by huge quantities of volcanic ash pumped into the atmosphere. Is the stage being set for rapid increases in volcanic activity both in the Torrid Zone and in other vulnerable spots at the earth's surface?

We know that a number of Torrid Zone volcanoes, lying between the tropics of Cancer and Capricorn, are already acting up, but how many of us know that the world's *most* active Torrid Zone volcano, Loihi (pronounced low-EE-hee), is in the United States. It is a growing, two-mile-high underwater volcano southeast of the southern tip of the island of Hawaii and has been in continuous eruption since 1982. In July 1996, 31 magnitude 4.0 and above quakes were recorded, prompting the Civil Defense director of Hawaii "to warn residents to head for higher ground immediately if they felt an earthquake. A quake of 6.8 could generate a local tidal wave in 15 minutes with no time for sirens or emergency broadcasts." (*New York Times*, 7/26/96)

And what is the most immediate volcanic hazard on the mainland? It appears to lie in the Mono Lake-Long Valley area of eastern California, currently the most active volcanic area in the continental United States.

The Long Valley caldera—the basin of a much older volcano, depressed and underground—that stretches 18 miles to the east of Mammoth Mountain—last erupted 250 years ago. But, "since 1980, it has produced nearly continuous swarms of earthquakes . . . a dome of hot magma is rising . . . [and] the land in the center of the caldera has risen two feet." (*New York Times*, 7/23/96) And, following several magnitude 6 quakes in 1989, up to 1,200 tons per day of carbon dioxide gas has been seeping out the flanks of Mammoth Mountain, poisoning soils and killing trees (whose roots need oxygen) and burrowing animals.

"About 700,000 years ago, a volcano at the [Long Valley] site expelled enormous quantities of incandescent ash . . . that traveled at speeds exceeding 100 miles per hour. So great was the volume of hot ash and so high its velocity that one arm of the ashflow surmounted the steep eastern face of the Sierra Nevada, a granite barrier thousands of feet above the erupting vents. Overtopping the mountain crest, the ashflows rushed west down the San Joaquin drainage, perhaps as far as the Central Valley of California" (S. Harris, 1990, *Agents of Chaos,* Mountain Press, Missoula).

Just to the north of Long Valley lie the Mono-Inyo craters. If an eruption is restricted to this part of the increasingly agitated area, it may resemble activity like that of A.D. 1350, when a chain of erupting vents produced widespread ash falls over central California and western Nevada and thick streams of lava that, if repeated today, would cause disruption of the communities all along U.S. 395, eastern California's mainline highway between Bishop and Bridgeport.

A final thought on possible volcanic fire to come leads us to the Yellowstone caldera. This volcano is underlain by a persistent hotspot generated by a thermal plume in the mantle.

Over the last two million years or so, the Yellowstone volcano erupted explosively three times, producing some of the greatest eruptions in Earth's history. These eruptions seem to have been spaced roughly 600,000 years apart, and it has been this length of time since the last one. The third (last) climactic eruption is believed to have ejected at least 250 cubic miles of new magma, or roughly a thousand times more than the eruption of Mt. St. Helens in 1980. The enormous ash canopy that was produced probably darkened skies over much of North America because Yellowstone ash layers are found in Kansas, Saskatchewan, and California. A repeat of such an event would perhaps fit in with reading 3976-15, where we find:

As to the material changes that are to be as an omen, as a sign to those that this [spiritual, mental, and physical change coming to Earth, presumably beginning in 1998] is shortly to come to pass—as has been given of old, the sun will be darkened and the earth shall be broken up in divers places ...

PART TWO

ARE THE PREDICTIONS CREDIBLE?

3

Have the Earth Changes Begun?

As the critical 1958-to-1998 period draws to a close, is there any credible evidence that the predicted geophysical changes are beginning to occur? We'll consider two lines of evidence. The first has to do with predicted "breaking up" in the South Pacific and associated sinking or rising of the crust in the Mediterranean area on the opposite side of the planet. The second concerns a sudden geophysical change in northern Europe.

In 1932, Cayce was asked:

Q How soon will the changes in the Earth's activity begin to be apparent?

A When there is the first breaking up of some

conditions in the South Sea (that's South Pacific, to be sure), and those as apparent in the sinking or rising of that that's *almost* opposite same, or in the Mediterranean, and the Aetna [Etna] area, then we may know it has begun. [Emphasis added.]

Q How long before this will begin?

A The indications are that some of these have already begun, yet others would say these are only temporary. We would say they have begun. '36 will see the greater changes apparent, to be sure.

311-8

April 9, 1932

The reference to " '36 will see the greater changes apparent" can be viewed in the same way that the subject of Earth changes in 1936 was dealt with in Chapter One; that is, the main geophysical events of 1936 were stated in reading 5748-6 to be "the upheavals in the interior of the Earth." And the phrase, *the greater changes apparent*, probably refers to events in the interior of the Earth that would be apparent to the psychic source that communicated reading 5748-6, and also, perhaps, 311-8. To be apparent at the Earth's *surface* would imply perception of observable shifts in the rates and/or intensities of ongoing seismic or tectonic activities. It would take a lengthy literature search to detect any such shifts in surficial seismotectonic patterns as having occurred in 1936. A cursory check shows nothing out of the ordinary. We'll move on then to consideration of evidence for "sinking or rising" near Mt. Etna, as apparent at the Earth's surface as we approach the close of the 1958-1998 period during which the Earth changes are scheduled to *begin* to occur.

Italy and Sicily are situated in the central Mediterranean region within the boundary zone between the African and Eurasian crustal plates. (See Fig. 5.) Strong, deep-focus earthquakes are known from this tectonically active zone. Mt. Etna, on the east coast of Sicily, is the highest volcano in Europe. Etna's large volcanic cone stands in a great Pliocene-age subsidence bay of the Ionian sea. During historical times it has erupted about every six years. But here at the end of the 1958-1998 period we find a marked increase in volcanic activity. Etna's 1991-1993 eruption represents the most significant event in the past three centuries, both in terms of duration (472 days) and total volume of erupted lava, and it has produced marked ground deformation as well (A. Bonaccorso, 3/1/96, *Geophysical Research Letters*, v. 23, p. 451). Since 1993, the pace of activity has resumed with violent eruptive activity from Etna's northeast crater, activity that was increasing into January of 1996. And, as of mid-1996, Etna and Stomboli were experiencing ongoing intense activity.

Let us continue to examine recent trends in sinking and rising in the Mediterranean in the area around Mt. Etna. Reading 311-8 says that one signal of the beginning of changes in the Earth's activity will be "sinking or rising of that that's *almost* opposite [the South Sea], or in the Mediterranean, *and* [emphasis added] the Etna area . . ."

The Mediterranean basin has a long history of instability marked by powerful eruptions. The island of Thera (also spelled Thira and sometimes called Santorin or Santorini), the most southerly island in the archipelago of the Cyclades in southern Greece, was almost blown away in the most powerful eruption in recorded history about 3,500 years ago. The eruption destroyed the Minoan civilization. Thera was once about ten miles in diameter, perhaps 5,200 feet high, and was known as "the very beautiful island." G.M. Friedman, in *Geotimes*

(March 1992), wrote: "The volcanic explosion at Thera/ Santorini in the eastern Mediterranean, from which tephra was blown across the Mediterranean, and the uplift along the eastern margin of the Sinai peninsula, may both relate to events that originated deep in the mantle. Because these two major paroxysms coincide approximately with the timing of the exodus of the Israelites from Egypt, actual geologic events may be factors in the background of the biblical account."

Friedman points out that tephra or ash emitted from volcanic explosions can spread widely from its source and influence climate. Tephra from the massive explosion of Thera rained not only on much of the eastern Mediterranean, but it was carried by the winds around the globe and had a deleterious impact on agriculture due to changes in climatic conditions. We know this because records of that period from China describe dimming of the sun, crop failure, and famine. The Thera eruption is recorded in ice cores from Greenland.

Current maps of seismic risk for Greece (S.C. Stiros, *EOS*, v. 76, Dec. 12, 1995) show that seismic coastal uplifts occur along the east coast of Greece, from Euboea northward to Stomoin, and along the southwest coast of Crete. Southwestern Crete was seismically uplifted by up to 30 feet around 360 A.D. Remains of shorelines along the nearly tideless coasts of the Aegean Sea testify to a relative sea-level drop of around 3.3 feet in Euboea about 1000 B.C. and 400 B.C., and in Thessaly at about 400 A.D. Stiros presents evidence that these fossil shorelines are indicative of seismic uplifts and not climatic changes of sea level. Occasional large shocks at intermediate depth occur in the Aegean region, some of which have caused much damage in Egypt, southern Greece, Crete, and other Aegean islands. Thus, should Cayce-readings-style Earth changes begin, we should expect Greece to become rather tectonically upset, and inhospitable to human life. Maximum seismic risk in Greece currently coincides with the Ionian Islands, off of the west-central part of the country.

Eastern Sicily and far southwestern and central Italy seem quite vulnerable. On the southern shore of the Mediterranean, much of northernmost Tunisia, Algeria, and Morocco falls into moderate to high-risk seismic zones. The geological instability of portions of the Mediterranean basin cause us to consider the basin at high risk when the significant Earth changes begin. These instabilities may be expected to respond quickly and significantly when the mantle starts moving in concert with a pole shift.

Much of the coast of Israel is tectonically very active. Recent deposits of marine shells have been found at elevations up to 130 feet above sea level. In several locations, these beds overlie ancient sites dating from the time of the Roman conquests to the period of the Crusades. Also, undisturbed Roman masonry (dating to about 10 B.C.) with lead-cast joints has been found more than 30 feet *below* sea level, giving evidence of subsidence of an ancient harbor by at least 30 to 45 feet. These indications of very large vertical movements of land near the shoreline in historical times are well documented. One can only imagine what the future might hold for Israel, with its tectonically active coast on the west and a major strike-slip fault in the Dead Sea fault zone on the east.

But what of "sinking and rising" of the Earth's crust that will become "apparent" in the Etna area, beyond the volcanic mountain itself, as an indicator of forthcoming "changes in the Earth's activity"? Two papers published in the book, *Volcano Instability on the Earth and Other Planets* (Geological Society Special Publication No. 110,

1996), provide important information. In the first, "Coastal Elevation Changes in Eastern Sicily: Implications for Volcano Instability at Mount Etna," C. Firth and others write that radiocarbon-dated remains indicate that the volcanic coast of Sicily, from Catania to Capo Schiso, and the northeastern shoreline, from Taormina to Milazzo, "have been uplifted at a rate exceeding 1.5 mm per year during Holocene times (roughly the last 10,000 years), *although more recent rates of uplift may have been greater*" (emphasis added). The authors attribute this emergence to "a regionally uplifting sub-volcanic basement"; that is, the Earth's crust is increasingly bulging upward in the area to the north and east of Mt. Etna.

The second paper, entitled "Recent Uplift of Ischia, Southern Italy," by S. Alessandro and P. E. Imbriani, discusses the volcanic island of Ischia, off Naples, that has long been tectonically active. These authors state that "historical evidence suggests that during the last two millennia parts of the north coast were submerged by up to 4 m[eters] whereas the southern coast underwent as much as 25 m [82 feet] of emergence." Finally, R. Westaway, in a journal paper entitled "Quaternary Uplift of Southern Italy," documents present-day regional uplift of the western portion of southern Calabria, Italy, of 1.67 mm per year, about 1.0 mm of which is interpreted to be due to regional uplift. The western margin of the Calabria area is about 50 miles northeast of Mt. Etna, across the Strait of Messina from Sicily. These three studies make *apparent* to us "the sinking or rising of that . . . in the Mediterranean, and the Etna area" mentioned in reading 311-8 above.

Is there a similarly unstable area of the crust "*almost* opposite," in the South Pacific. An area subject to "breaking up"? *Directly* opposite the Mediterranean region would refer us to an oceanic tract bounded roughly by longitudes 146° to 180°W, and latitudes 32° to 42°S. This tract *almost* includes the whole of the extremely tectonically active Kermadec-Tonga Trench (see KTT in Fig. 5), northeast of New Zealand.

It is here that we find the Earth's most active zone of *mantle* seismicity. The mantle seismicity arises from the subduction of the Pacific plate at the trench, and about two-thirds of global deep quakes are located here. In 1995 several geophysicists studying the area wrote:

> It is not known why this slab [Pacific plate] generates so many more earthquakes than other subducting slabs worldwide. Above the subduction zone the active Tofua (Tonga) volcanic arc is separated by the V-shaped Lau basin from a remnant arc, the Lau ridge, located at the eastern edge of the Australian plate. The subduction rates in Tonga result from a combination of Australian-Pacific plate convergence and the opening of the Lau basin (Australian—Tonga divergence). The intense mantle seismicity beneath Tonga—Lau may arise because this area has the highest subduction and back-arc extension rates on Earth.[2]
>
> The March 9, 1994, Fiji earthquake (Fiji = 7.5M) suggests a new mode of seismic deformation is active in the northernmost termination of the Tonga-Fiji slab, cutting across the dense cluster of seismicity of the last 30 years.[3]

We marvel at the vision revealed in the readings in 1932, to have described the breaking up of an area of the South Pacific "almost opposite" to the Mediterranean/Etna area some three decades before scientists even began to understand the significance of the Kermadec-

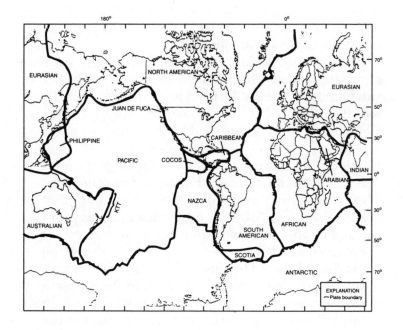

Figure 5
Earth's crust is broken into several large and numerous small fragments or lithospheric plates. Because these fragments are constantly being moved in various directions by tectonic forces in the underlying mantle, the fragments are often referred to as tectonic plates. Earthquakes and volcanoes occur most frequently along plate boundaries. The letters KTT northeast of New Zealand indicate the location of the Kermadec-Tonga Trench.

Tonga Trench (KTT) to global tectonics.

In addition to considering the Tonga trench and the Tonga-Fiji slab to be the most likely area for Cayce's "breaking up" in the South Pacific, there is another area of significant crustal instability *"almost"* on the opposite side of the globe from the Mediterranean and Mt. Etna. This is the Vanuatu area at the northern end of the New Hebrides Trench. Here, volcanologists recently found[4] that the entire sea floor between the islands of Tongoa and Epi was "a seven-mile-wide crater—the remnant of a gigantic eruption as powerful as 2 million Hiroshima-type atomic bombs. Vegetation charred during the eruption . . . [has been] . . . dated to between 1420 and 1475." Geophysicists (Prevot, et al.) investigating the shallow, double-seismic zone near Vanuatu in the New Hebrides island arc have found evidence for the fragmentation ("breaking up") of the descending Australian plate there,[5] while another team (Taylor, et al.) asserts that "if eastward displacement of the central New Hebrides persists . . . then the arc is destined to undergo extensive fragmentation . . . "[5]

Thus the readings, over sixty years ago, described roughly antipodal regions of the crust and mantle that are significantly unstable, and likely to be the first to be sensitive to global "changes in the Earth's activity." The most visible of the two opposing regions is the Mediterranean area, including especially Mt. Etna in northeastern Sicily and the Callabria, Italy, area. But modern-day seismological monitoring being what it is, any breaking up of conditions in the crust and mantle beneath the waters of the South Pacific will be documented as it accelerates. The KTT and Tonga-Fiji-slab area in particular bear careful watching in the days ahead. Intensification of seismotectonic activities on either side of the Earth will provide mankind with an excellent early-warning system for more significant Earth changes to come.

Finally, we must also note that Mururoa, site of France's nuclear weapons tests, is also located within range of the area in the South Pacific that is "almost opposite" the Mediterranean area. Mururoa is about as near to our target area as are the New Hebrides Islands, but *not* as close to it as is the Kermadec-Tonga Trench. The volcanic base of the Mururoa atolls has been broken up and weakened by France's repetitive nuclear tests and may leak radioactive wastes for the next millennium. Already, radioactive iodine, cesium, and plutonium have been found outside the 12-mile limit from the Mururoa lagoon. France conducted a nuclear test there as recently as January 1996.

Now what about the possibility of a sudden geophysical change of some sort in Europe? Reading 3976-15 says:

> **The upper portion of Europe will be changed as in the twinkling of an eye.**
>
> **January 19, 1934**

In the 1959 edition of *Earth Changes: Past-Present-Future,* I considered two possible causes of such a change. The first described geological evidence for long-term uplift of Scandinavia since melting of the last ice sheet there and suggested that this uplift might suddenly accelerate. This would perhaps agree with reading 3976-15 that says there will be "upheavals in the Arctic and in the Antarctic." The second possible cause involved sudden blockage or diversion of the Gulf Stream that brings warmth to Europe. When that was written, oceanographers had not yet worked out just how heat is carried into the North Atlantic. The Gulf Stream's warmth is only part of the picture.

In the last 35 years much has been learned about the

ocean-atmosphere-climate system. Cayce was way ahead of scientists when he said:

> **... there will be for the next few years many changes of an exceptional nature in [atmospheric] heat and cold. Not by radiation nor by the courses changed in the sea; for the heat is as the reflection from that created in the Earth's plane, rather than that administered. Only in the relations to each other would these be considered ...**
>
> **Q Is Herbert Janvrin Browne's theory correct—whereby weather is forecasted several years in advance by measuring solar radiation and its action on the ocean currents?**
>
> **A Were these varied accounts considered of that information intimated here, these would be *not* correct, for this may be established as a theory: that thrown off will be returned. As the heat or cold in the various parts of the Earth is radiated off, and correlated with reflection in the Earth's atmosphere, and this in its action changes the currents or streams in the ocean; and the waters bring or carry the heat in a manner to the various shores, or bring cold or carry cold to the various shores.**
>
> **195-29**
> **May 28, 1926**

The part of the theory advanced in the reading that refers to heat being radiated from the Earth and then correlated with "reflection in the Earth's atmosphere" is essentially the current carbon-dioxide theory of atmospheric heating that leads to slow climate warming (the "greenhouse" effect). Both atmospheric carbon dioxide and methane absorb infrared radiation thrown off from the Earth's surface as it is heated by the sun. This heat is returned (reradiated downward), increasing the temperature of the lower atmosphere and of the Earth's surface, including the oceans. Cayce was about 30 years ahead of atmospheric scientists when this reading was given in 1926.

Among the best bets for early detection of global climate change is documentation of progressively warmer ocean water. Although not terribly well expressed relative to today's scientific terminology, the theory propounded in the reading has clear overtones with the recent work of oceanographers who look at how the ocean and atmosphere have interacted in the past as climate swung from one extreme to another, sometimes with stunning abruptness. W. S. Broecker, for example, wonders why we aren't more concerned about our immediate future, because as the greenhouse warming pushes the climate system to temperatures unheard of for millions of years, changes might happen all too quickly. In a recent article,[6] he says, "Global temperatures have been known to change substantially in only a decade or two. Could another jump be in the offing?"

Professor Broecker has detailed a major reorganization of the climate about 135,000 years ago. It involved creation of a conveyor belt of ocean currents carrying heat or cold nearly around the globe and to the shores of the various continents. (See Fig. 6.) Thus, with respect to the matter of the origin of Earth's climates, there is really very little difference between what the 1926 reading said and what scientists like Broecker are saying today.

As for the climate immediately ahead, Broecker is saying that sudden warming in the North Atlantic Basin is the next most likely event. He recently commented on studies of Greenland ice cores that show that the northern Atlantic basin was affected by rapid climate changes over the past 100,000 years or so, due to switches in the

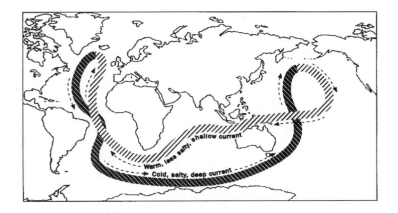

Figure 6
Fresh water and heat added in the Pacific are carried into the Atlantic by shallow currents. In the North Atlantic, heat and fresh water are removed by cold air from Canada. This results in an increase of water density, causing it to sink. Europe is warmed when the conveyor carries a large heat load, as now, and leaves that area in the cold if the conveyor slows or stops. (Adapted from W.S. Broecker and G.H. Denton, *Scientific American,* Jan. 1990, p. 53.)

mode of operation of the Atlantic Ocean's thermohaline circulation. (This is a much smaller-scale circulation than that described in the conveyor-belt conceptualization; it is driven by relatively local fluxes of heat and fresh water through the ocean's surface.) He says that these switches in ocean circulation would be "responsible for large temperature changes in the land surrounding the northern Atlantic."

Of particular interest to the sudden warming speculation are any observations of changes in the heat load of the upper levels of the Atlantic Ocean that are flowing from the south, northward toward the North Atlantic Basin. In 1994, a group of Spanish oceanographers published a study[7] that compared water temperatures measured in 1957, 1981, and 1992 throughout a cross section of the ocean between Africa and Florida, along 24° N latitude. They concluded that the waters there "had warmed appreciably down to 3,000 meters depth."

We started this section by suggesting that the reading that said that "the upper portion of Europe will be changed in the twinkling of an eye" might be referring, not to crustal changes, but to sudden warming of the northern Atlantic basin. This warming would be due either to switches in the mode of operation of the northern Atlantic ocean's thermohaline circulation or to an increased heat load in the oceanic "conveyor belt" that carries heat or cold nearly around the globe. Both mechanisms may come into play.

Is there, then, any evidence that such unexpected warmth is showing up in the North Atlantic Basin? Quite possibly there is. Consider the 1994 report that describes signs of warming evident in parts of Spitzbergen, 600 miles north of Norway and about 250 miles south of the boundary of permanent Arctic pack ice. The leader of a team of scientists studying insects there for the period 1990-1993 has reported that "the general trend seems to be that the western coast of Spitsbergen certainly is getting warmer. Last year [1993] it was very warm for quite a considerable period early in the year—warmer than you'd expect even from the extremes of the long-term averages." Climate records for the area suggest these feelings are well founded. "Over the past few decades," continues the article, "the overall trend in mean spring and summer temperatures has been upwards. Whether this reflects genuine climate change or is simply part of a broad natural cycle is not yet clear."

Again, this apparent trend in climate change is in the direction of the prediction of the readings, assuming that "the upper portion of Europe will be changed as in the twinkling of an eye" refers to sudden climate warming. And clearly, there would have to be such warming for the prediction of reading 3976-15 (1934) to come true. This reading says that, beginning in the period 1958-1998, "open waters [will] appear in the northern portions of Greenland." For this to happen there must be enough warm water in the North Atlantic Basin to melt the permanent pack ice that currently exists off northernmost Greenland. Interestingly, a recent publication describes a record minimum in Arctic sea ice cover during 1990, attributing the 1990 situation to being part of a larger-scale temperature-anomaly pattern.[9] And satellite surveys show that the reduction of Arctic sea ice accelerated between 1978 and 1994.[10]

One might argue that the seismotectonic and climatologic phenomena cited above represent no more than newly discovered *ongoing* geophysical processes, and that they are, therefore, not presumptive evidence of the beginning of Cayce-readings-type Earth changes. I would argue, however, that inasmuch as the Cayce readings saw these geophysical phenomena as emblematic of the beginning of the period of accelerated Earth changes, *before the importance or in some cases even before the existence of these particular phenomena were recognized by scientists*, we can just as well assume that the readings are also correct that these phenomena indicate that the period of upheavals and pole shift is upon us. I concur, then, with the words of reading 311-8: "We would say they [the Earth changes] have begun."

4

Is America Vulnerable?

Certain of the readings were explicit in predicting changes in Cayce's native land. So let's look at those locations mentioned specifically as having a built-in potential for dramatic change—crustal uplift or subsidence and destruction by earthquakes or volcanic eruptions. Do geophysical trends offer any evidence to lend credence to these dire forecasts? And what of present societal trends that, according to the metaphysics of the readings, may determine whether, or to what degree, the predicted changes will occur?

We must not shrink from the realization that, geophysically speaking, we exist on a dynamic, changing planet. It's been that way from Earth's birth and will continue to be so millions of years hence. This understanding is implicit in the Cayce readings, but where the readings differ from orthodox geoscience is in the number, scope, and rapidity of geophysical changes predicted for the next several years.

THE WESTERN STATES

We return now to reading 3976-15, given in 1934:

> **As to the changes physical again: The Earth will be broken up in the western portion of America ... The Earth will be broken up in many places. The early portion will see a change in the physical aspect of the west coast of America.**

To this we will add the following extract of a reading given in 1936:

> **Q What is the primary cause of earthquakes? Will San Francisco suffer from such a catastrophe this year? If so, give date, time, and information for the guidance of this body, who has personal property, records and a wife, all of which it wishes safety.**
> **A We do not find that this particular district (San Francisco) in the present year will suffer the great *material* damages that *have* been experienced heretofore. While portions of the country will be affected, we find these will be farther *east* than San Francisco—or those *south*, where there has not been heretofore the greater activity.**
> **The causes of these, of course, are the movements about the Earth; that is, internally—and the cosmic activity or influence of other planetary**

forces and stars and their relationships produce or bring about the activities of the elementals of the Earth; that is, the Earth, the Air, the Fire, the Water—and those combinations make for the replacements in the various activities.

270-35
January 21, 1936

Finally, we have the following from a 1941 reading:

As to conditions in the geography of the world, of the country changes here are gradually coming about ... For, many portions of the east coast will be disturbed, as well as many portions of the west coast, as well as the central portion of the U.S.

1152-11
August 13, 1941

The statement in reading 270-35 to the effect that earthquakes are caused by internal movements "about the Earth" finds support in recent work by O. Cadek, et al., who determined that dynamic processes up to 1,000 km deep in the mantle (see Fig. 2) can be linked with surface tectonic activity, which would include earthquakes (*Earth and Planetary Science Letters*, 1995, v. 136, p. 615).

With regard to the predicted locations of earthquakes in the vicinity of San Francisco, in 1936, no strong shock was felt in all of California in 1936. Published records indicate that all of the eight principal shocks in California and Nevada, for the period 1936 to 1950, were to the southeast of San Francisco.

A 1960 article in the respected British periodical *Nature* on the concept of "planetary forces" as an element to be considered in the mechanism of earthquake generation said this:

To state the plain facts: one is led to the conclusion that the position of Uranus within 15° of the meridian at the moment of great earthquakes can be regarded as significant and that there exist times of longer period (several years) when it is very highly significant. It is quite obvious that the strains and stresses in the Earth's crust are the primary cause of earthquakes, but it has been shown ... that the *timing* of the event can be described by the position of Uranus. If it is necessary to explain this within the limits of present-day science, attention would be directed to the fact that Uranus is the only planet of which the direction of its axis of rotation coincides with the plane of its orbital revolution. A possible magnetic field would influence the solar plasma in a way quite different from all other planets. The earthquakes which destroyed Agadir ... occurred with Uranus only 4° from the meridian. Anybody in Agadir, knowing of my communication in *Nature* ... would have kept away from buildings at the time of Uranus being near the meridian of Agadir which was from about 10 hrs. to 12 hrs. A.M. or P.M. The destruction of the town occurred at 11 hrs. P.M., local time. An unbiased approach to these problems, of which the correlations of Uranus are only a part and a first step, may help humanity. (R. Tomascheck, *Nature,* 1960, v. 186, p. 338.)

Los Angeles and Coastal Southern California

In 1941, Edgar Cayce was asked:

Q Will Los Angeles be safe?

A Los Angeles, San Francisco, most all of these will be among those that will be destroyed before New York even.

1152-11
August 13, 1941

If there are the greater activities in the Vesuvius, or Pelee, then the southern coast of California—and the areas between Salt Lake and the southern portions of Nevada—may expect, within the three months following same, an inundation by the earthquakes. But these, as we find, are to be more in the Southern than in the Northern Hemisphere.

270-35
January 21, 1936

California's San Andreas fault and its system of subparallel faults are recognized as being the boundary zone between the Pacific and North American tectonic plates. These slabs of crust are constantly grinding against one another. There's no question but that recently there has been a sharp increase in quakes in the Los Angeles Basin relative to earthquake activity of previous decades. (See Fig. 7.) The trend concerns seismologists because it resembles the cycle of quakes that preceded the great 1906 earthquake in San Francisco. But some say that the rate of seismic release on the Los Angeles Basin faults, like the one that produced the Northridge quake of January 17, 1994, was just an example of a change in *style* of seismic release between two types of fault systems.[1] Note, however, that as suggested by reading 270-35 above, style is not our immediate concern; rather, it is a sudden acceleration of plate grinding that would "inundate" the Los Angeles Basin with earthquakes, involving whatever fault systems are incident to the type of plate interactions that might occur. There is no precedent in the geophysical literature to guide us here. We are left only with general speculations about how significant eruptions of Vesuvius or Pelée could relate, sometime over the following three months, to a sudden increase in earthquakes in California, Utah, and Nevada.

Certainly, this up-to-three-month warning could be quite useful. Just think about this fact: The quake caused by movement along the small fault beneath Northridge, the first quake since 1933 to strike directly under an urban area of the United States, produced the greatest fi-

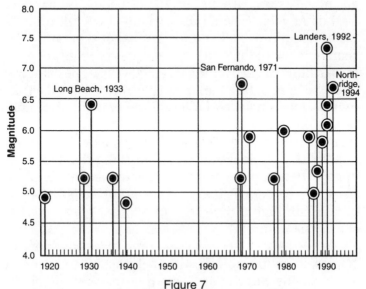

Figure 7
Major southern California earthquakes over the last 75 years.

nancial loss ($13 to $20 billion) from a natural disaster since 1906.[2] Then what if the main San Andreas fault just east of Los Angeles were to experience a big slippage? If it did so tomorrow, 10 million or more people in the developed areas of southern California could be affected. The big-quake recurrence interval on the San Andreas fault has averaged about 100 years at Wrightwood, California, and the elapsed time since the last big quake there (in 1857) has already been 135 years.[3] Since 1857, the fault in southern California has remained quiet, storing stress that could be released in a series of magnitude 7 quakes or one big magnitude 8+ shock.

Because proximity makes all the difference in earthquakes, the most affected people from a big San Andreas fault quake would be the 3 million residents who live astride the fault in San Bernadino and Riverside; but in Los Angeles, the most intense shaking would come from earthquakes beneath the city. In the worst-case scenario for Los Angeles, several thrust faults beneath the L.A. Basin might break all at once instead of separately. Researchers "warn that the L.A. Basin has probably been in an earthquake lull over the past 200 years. The calm will likely be broken by a barrage of Northridge-sized earthquakes, by a single, far larger quake, or some combination of the two."[4]

Again, what of the warning for coastal California? Reading 270-35 above implies a very long-distance connection between seismotectonic activities in one place (Mts. Vesuvius or Pelée) and resultant activities—expressed through a time lag—in another place (coastal southern California).

Mt. Vesuvius is located about nine miles southeast of Naples, in southern Italy. Although the last eruption of Mt. Vesuvius was in 1944—eight years after the above reading was given—the eruption did not involve truly "greater activities," by comparison, say, to the burial of Pompeii in 79 A.D. or the devastating eruptions of 1631, 1794, 1872, and 1906. At present there are about one million people at risk in the Vesuvius area, considered to be one of the most dangerous volcanic areas on Earth.

It is interesting to note in passing that Mt. Vesuvius lies within the remains of a much larger ancient volcano, designated Monte Somma. The end of the constructional phase of Monte Somma occurred 18,000 years ago as a plinian eruption (A. Bertagnini and P. Landi, 1995, *AGU Fall Meeting Abstracts*, p. F676), which is an explosive eruption in which a steady stream of magma and gas is expelled at high velocity from a vent. Since this was the final constructional phase of the great volcano, it is interesting to note that it was probably being built for at least a thousand years prior to the 18,000-year-old date. This means that its building coincides with the destruction of the main part of Atlantis, some 2,300 miles to the west. Reading 364-11 says that this event occurred 19,400 years ago and involved volcanism and the sinking of a portion of the continent of Atlantis.

Mount Pelée, not to be confused with Pele, the Hawaiian fire goddess, is located on the northern coast of the island of Martinique in the West Indies. It erupted in 1792 and 1851. In May 1902, it suddenly erupted again when masses of lava, pumice, and hot ash burst from the south side of the 4,600-foot-high mountain. In one day, 30,000 people died. That same year, 15,000 people lost their lives to the eruption of Soufriere volcano on the nearby island of St. Vincent.

Is there anything going on volcanologically in the Caribbean area at this writing (mid-1996) that might be signaling that renewed eruptive activity is in store for Mt. Pelée, here at the end of the 1958-1998 period that reading 3976-15 said would be the beginning of extraordinary Earth changes ahead? Perhaps. We note that volcanoes in Costa Rica and Nicaragua have been showing increased activity, near the western end of the Caribbean plate. Of greater significance is the eruption of the Soufriere Hills volcano on the island of Montserrat, only 150 miles north-northwest of Mt. Pelée. This volcano had been dormant for about 16,000 years, when it suddenly began rumbling and spewing ash on July 18, 1995. In April 1996, near-continuous seismic activity began to occur, and the volcano started belching ash clouds as high as 7,000 feet. This activity on Montserrat sounds similar to the "premonitory" eruption of Mt. Soufriere, on St. Vincent, which preceded the cataclysmic eruption of Mt. Pelée in 1902.

The last eruption of Mt. Pelée was in 1929; it remained active for three years but no lives were lost from the eruptions. Just prior to Pelée's May 8, 1902, eruption "an earthquake shook Quezaltenago in Guatemala, 2,100 miles away; on May 7 a major volcano erupted at Soufriere, on St. Vincent, only 100 miles away; and on May 10, the Izalco volcano erupted in El Salvador, 1,938 miles away. These four events, occurring within days of each other—fractions of seconds in geologic time—provide strong evidence of the interrelationship between earthquakes and volcanoes (especially when we realize that these locations are on opposite ends of the Caribbean [lithospheric] plate)" (L. Levy and M. Salvadori, 1995, *Why the Earth Quakes*, W.W. Norton & Co.).

One can conjecture that once pole shift begins, and the tectonic plates and mantle bulges get moving, Vesuvius and Pelée, due to their sensitive locations, will be among the first to respond. Their eruptions will then be followed by numerous seismotectonic events in the Southern Hemisphere, and, additionally, areas of coastal California and parts of Nevada and Utah as well.

The San Diego Area

Much of the city of San Diego, including the business center and harbor, is situated on a low sandy area that could be subjected to intensity VIII shaking. Although no magnitude 5.5 or greater quakes have occurred close to San Diego for the last 80 years,[1] the San Andreas fault lies only 65 or so miles east of the city. However, in line with the importance of proximity to quakes mentioned above, the greater threat to San Diego may be related to the Rose Canyon fault zone that recently has been found to project into the downtown area. "Preliminary interpretations of [fault] exposures and radiocarbon-age determinations suggest active faulting and possibly a recent ground rupture event" along the fault zone there.[5] The Rose Canyon Fault extends northwestward offshore, reappearing as the Newport-Inglewood fault that runs directly under Los Angeles. Movement on this front caused the 1933 (M 6.3) Long Beach quake.

The San Francisco Bay Area and Northern California

The San Francisco Bay Area was far more seismically active in the nineteenth century, before the 1906 quake, than it has been since. But things began picking up in the Bay area about 30 years ago.[6] After the Loma Prieta quake that damaged San Francisco and Oakland in October, 1990, a U. S. Geological Survey working group estimated that there is a 67 percent chance that one of the major faults in the region will unleash a magnitude 7 or larger quake in the next 30 years. A quake this size under the populated Bay area would wreak significantly more damage than the 7.1 Loma Prieta shock that originated in the rural Santa Cruz Mountains. (See Fig. 8.)

Scientists are concerned that the devastation to the

San Francisco Bay Area caused by the Loma Prieta quake may echo the high earthquake activity that rocked the same area during the nineteenth century. There are two equally disturbing possibilities. "One is that an historical pattern of large paired quakes—involving faults on the San Francisco (Peninsula) and the Oakland (East Bay) sides of the Bay—may be about to repeat itself."[7] If so, the East Bay would be the next site of a large, but not enormous, shock, probably on the Calaveras or Hayward faults. The Hayward fault cuts through the cities of Fremont, Hayward, Oakland, and Berkeley. According to a state study, a magnitude 7.5 event, the largest credible quake on the Hayward, might kill up to 4,500 persons, injure more than 50,000, and wreak social havoc among the 5 million residents of the area.

The other possibility is that the stress between tectonic plates on the Peninsula side has now been increased significantly by the slippage in the Loma Prieta area to the south, requiring the part of the San Andreas fault (Fig. 8) that runs toward San Francisco to be the next to move. The San Andreas segment on the Peninsula has been locked tight since 1906, as had the Loma Prieta segment until 1990. It runs northward by Santa Clara, Palo Alto, and San Mateo to San Francisco itself. What is quite discomforting is that in the nineteenth century two Hayward-fault shocks of about magnitude 6.7 were matched within two to three years by ones of magnitude 6.5 to 7.0 on the San Francisco side of the Bay, along the San Andreas.

The following reading was given for a woman born just north of San Francisco, in either Santa Rosa or nearby Petaluma (see Fig. 8).

Before this we find the entity in that land now known as the American, during the periods when

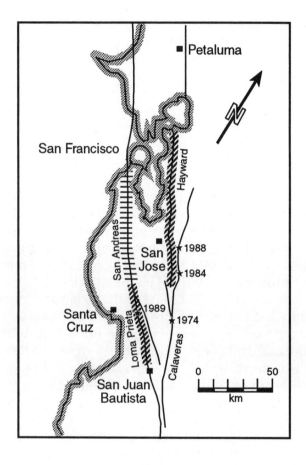

Figure 8
The San Francisco Bay Area is underlain by faults that were quiet after
the great 1906 quake but began to come to life in the 1950s.

the Lemurian or the lands of Mu or Zu were being
in their turmoils for destruction.

And the entity was among those that—in what
is now not far from that land in which the entity in
this sojourn first saw the light—(that must in the
near future fade again into those joinings with the
land of Mu)—established a temple of worship for
those that escaped from the turmoils of the shifting of the Earth at that particular period.

509-1
February 5, 1934

Here we have an intimation that *in the near future* the
Santa Rosa/Petaluma area will be "fading" into Mu or
joining with the Pacific lithospheric plate. The eventual-
ity of such a joining might not completely surprise geologists involved with reconstructing Earth's history by
application of the theory of plate tectonics.

Before discussing "joinings with the land of Mu," we
need to realize that California came into being as a result of (1) the repeated plastering-on of fragments of
continental and oceanic crust transported mostly from
south-to-north or west-to-east by movements of the Pacific lithospheric plate, or (2) by underthrusting of the
North American plate by the Pacific plate. Some of the
details of formation of the coast range of mountains and
the Napa Valley, in the Santa Rosa and Petaluma area,
are rendered almost poetically in this passage from John
McPhee's *Assembling California:*[8]

As we go up the stream valley and arrive at the
shore of Lake Berryessa [30 miles east of Santa
Rosa], we pass through huge road-cuts of sedimentary rock whose bedding planes, originally
horizontal, have been bent almost ninety degrees

and are nearly vertical. Reaching for the sky in distinct unrumpled stripes, the rock ends in hogbacks, jagged ridges ... these are the bottom layers of the Great Valley Sequence ... some of the strata that were folded against the Franciscan melange when it rose (or was pushed) to the surface as the latest addition to the western end of the continent.

Today, terranes like that just described—joined to the continent by ancient suturing processes—are being sliced off again within the west coast's active tectonic strip and are being conveyed northwestward by displacements related to interactions between the Pacific and North American plates.[9] The most visible example of this is found where Baja California is being transferred from the North American to the Pacific plate. "The transfer of the Baja peninsula to the Pacific plate during the eastward shift of the plate boundary is generally assumed to have been completed by 3.6 Ma [millions of years before the present], when sea-floor spreading commenced along the Gulf rise in the southern Gulf of California."[10]

The above fragment of reading 509-1, in concert with several other readings, may be interpreted to mean that the sudden acceleration in plate motions that will result from the upcoming pole shift will cause the transfer of the sliver of the west coast, west of the Hayward fault and its northwestward extension (Fig. 8), from the North American to the Pacific plate. This crustal sliver could well include the Santa Rosa/Petaluma area, because the Pacific-North American plate boundary for the lower crust and upper mantle is located approximately beneath the surface trace of the Hayward fault in the San Francisco Bay Area[11] and, by inference, beneath the trace of northwesterly extension of the fault (along the

Rodgers Creek fault) just to the east of Petaluma and
Santa Rosa (Fig. 8). J. Prims and K. P. Furlong described
the possible details of the future evolution of the San
Francisco Bay crustal block as it is being "transferred
onto the Pacific plate, becoming a new captured crustal
terrane with an associated piggyback basin (San Francisco Bay) on it."[11] This captured crustal terrane could
well include the Petaluma/Santa Rosa area and correlates nicely with the psychic vision of that area fading
"again into those joinings with the land of Mu" or the
Pacific lithospheric plate. As for the timing of the event,
we are left to muse on the "must in the near future" statement in reading 509-1.

We continue northward along the west coast now,
continuing to try to understand how an acceleration or
change in the nature of plate-tectonic activity implied
by the Cayce readings relates to the statement that "Los
Angeles, San Francisco, most all of these will be among
those that will be destroyed . . . " (1152-11 earlier). Consider Cape Mendocino, California (Fig. 9), for example,
where *three* tectonic plates come together in a "triple
junction"; the plates in question are the Pacific, Gorda,
and North American plates. Some of the highest rates of
crustal deformation, surface uplift, and seismic activity
in North America occur at this Mendocino triple junction. On April 25-26, 1992, for example, three powerful
quakes occurred here, along with 1.4 m of coastal emergence. According to the U. S. Geological Survey, the main
shock of the series "may be the forerunner of a much
more powerful, tsunami-generating earthquake in the
Pacific Northwest . . . Such an event could cause [great]
damage due to ground shaking, and the associated tsunami could devastate the coasts of California, Oregon,
Washington, and British Columbia."[12]

In early 1996 scientists detected a major volcanic upheaval at the north end of the sea-floor rift segment known as the Gorda Ridge (see Fig. 9). This is testimony to the ongoing geologic outpouring of material from the Earth's interior in this subsea region of the Pacific Northwest as heat is transported to the surface in the vicinity of the Gorda Ridge.

The Pacific Northwest

A recent study[13] concludes that the Northwest coast, from Vancouver Island, British Columbia, down to northern California, faces an earthquake hazard at least as great as that along the San Andreas fault. Earthquakes of magnitude 8.0 [M 8] or above are a virtual certainty every few hundred years, and "if the Cascadia subduction zone is storing elastic energy, a sequence of several great earthquakes [M 8] or a giant earthquake [M 9] would be necessary to fill this 1,200-kilometer long [seismic] gap . . . Strong ground motions from even larger earthquakes [M 8.5 up to 9.5] are estimated . . . and . . . if large subduction earthquakes occur in the Pacific Northwest, relatively strong shaking can be expected over a large region."[14] Seattle is about the same distance from the Cascadia rupture zone as is Anchorage from the Alaskan subduction zone. Anchorage was strongly shaken by a magnitude 9.2 earthquake in 1964.

Such earthquakes may also be accompanied by large local tsunamis. There have been estimates (K. Satake, 1994, for example) that an M 9 quake would cause tsunamis larger than 33 feet along the Pacific coast of northern California, Oregon, and Washington. Japanese scientists believe they have detected evidence for a Pacific Northwest megaquake that generated a tsunami that buffeted the Japanese coast three centuries ago.[15] Large *shallow* earthquakes can also generate tsunamis.

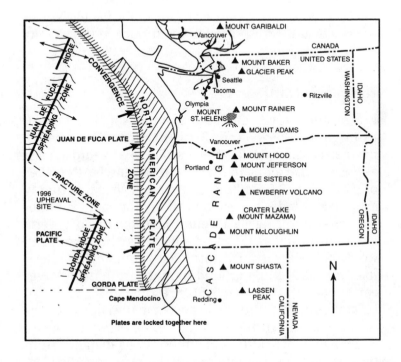

Figure 9
Pacific Northwest and associated offshore region. The position and
alignment of the Cascade Range volcanoes result from the slow
grinding together (convergence) of the Juan de Fuca and Gorda plates
with the North American plate as new volcanic rock is added at
spreading zones beneath the ocean. Arrows indicate relative movement
of the crustal plates, and shaded area shows where plates are locked
together, storing stress.

Such an historic quake was documented recently for the Seattle area; one that generated a tsunami in Puget Sound about a thousand years ago.[16]

Earthquake hazards in the Pacific Northwest or Cascadia, are divided into three different categories: those that occur along the coast as the subducting Juan de Fuca plate slips beneath the North American plate (Fig. 9); those that occur 40-66 km beneath the surface within the subducted plate; and those that occur in the crust, which is moving northward in a series of blocks bounded by faults. To these hazards we must add hazards related to tsunamis and volcanoes.

There are seven active volcanoes in the Cascade Range (Fig. 9), including Mt. St. Helens that erupted violently in 1980. The others are Mt. Baker, Glacier Peak, and Mt. Rainier, in Washington; Mt. Hood, in Oregon; and Mt. Shasta, Cinder Cone, and Lassen Peak, in California. As for volcanic hazards in Cascadia, consider that the eruption of Mt. St. Helens on May 18, 1980, was not a large eruption by world historical standards or even among prior Cascade eruptions (see Fig. 10). Other Cascade volcanoes are expected to erupt in the future. Some, like Mt. Rainier, could cause problems even without erupting. In the past, major sections of the mountain have simply collapsed, creating large avalanches and mudflows that swept through low-lying regions now home to 100,000 people. The U. S. Geological Survey's C. L. Driedger, who has studied the record of mudflows—or lahars—at Mt. Rainier, considers this volcano to be the most dangerous of the Cascade Range, in part because the mountain has a thick mantle of snow and ice that can melt to form floods and lahars. The growing population along the base of the volcano compounds this threat. The most recent giant lahar swept down off Rainier 200 years ago.

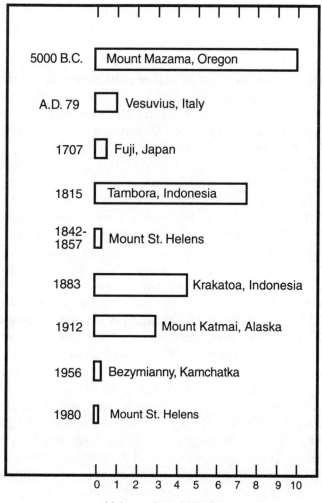

Volume, in cubic miles

Figure 10.
Comparison of volumes of ejecta thrown out by selected volcanoes during historic eruptions. The volume of ejecta from the May 18, 1980, eruption of Mount St. Helens is relatively small in comparison to amounts from several earlier eruptions, including those from Mount Mazama (Crater Lake) in Oregon.

Because the Cascade volcanoes have formed in response to movement of the Gorda and Juan de Fuca Plates beneath the North American Plate (see Fig. 9), as the plates move ever more quickly in response to pole shift, all of the volcanoes could begin to erupt simultaneously.

Salt Lake to Southern Nevada

Just what sensitive spots in the Earth's crust and mantle was Cayce's clairvoyant power revealing in 1936 when he said, *"If there are the greater activities in the Vesuvius, or Pelée, then the southern coast of California—and the areas between Salt Lake and the southern portions of Nevada—may expect, within the three months following same, an inundation by the earthquakes"*?

Geologists of 60 years ago had not yet put together the seismotectonic information for Utah, Nevada, and California to have made it possible, as some have suggested, for Cayce's mind to simply read the thoughts of geophysicists of that day to come up with his forecast. It was not until 1966, thirty years after Cayce's 1936 reading, that Alan Ryall, et al., published their ground-breaking maps of tectonic flux for the conterminous United States west of longitude 109°W.[17] "Tectonic flux" is a quantitative, mappable index of seismicity that is defined by a rather complex mathematical expression. Maps of such a flux represent a compromise between methods of illustrating seismic activity that stress numbers of shocks, such as epicenter maps, and methods involving the energy released by earthquakes, which tend to emphasize the largest events.

Figures 11 and 12 (based on figures 4 and 3, respectively, of Ryall, et al.) show that there are major fault zones in a belt trending southwestward from Salt Lake

into southern Nevada (Fig. 11), and that a belt of relatively high tectonic-flux coincides with most of these
fault zones (Fig 12.) And so, Cayce's psychic vision of a
likely seismotectonic belt within which an inundation of
quakes might take place conforms to the results of the
Ryall, et al., study. Note, too, that the seismotectonic belt
roughly coincides with important structural features like
the eastern limit of imbricate thrust sheets, part of the
boundary between the northern Basin and Range Province and the Colorado Plateau, and the axis of the Late
Cretaceous foreland basin.

The Wasatch fault zone (Fig. 13), at the north end of
the belt, straddles or lies adjacent to the Brigham City-
Salt Lake City-Ogden urban corridor that contains more
than 1 million people and great building wealth. It is
composed of at least 10 segments that have surface
traces 9 to 36 miles in length. Each segment has been
the source of repeated earthquakes, of about magnitude
6.5 to 7.5, occurring on the average every several hundred to several thousand years. "Potential losses to
buildings and life from shaking in future earthquakes in
the Salt Lake City region may reach $1 billion (1985 dol-
lars) or more, depending on the location of the earthquake relative to the urban area and the time of day the
earthquake occurs."[18] We note, too, that Tooele, Utah,
near Salt Lake, is home to America's largest stockpile of
chemical weapons.

In Cayce's "southern portions of Nevada," at the
southern end of the tectonic flux belt of Fig. 12 lies Las
Vegas, which has the largest population concentration
in the immediate region. The effects of "an inundation
by the earthquakes" (270-35) on the Las Vegas metropolitan area, nearby Hoover (Boulder) Dam, and the
nation's only high-level nuclear waste repository, now
under construction at Yucca Mountain, would be devas-

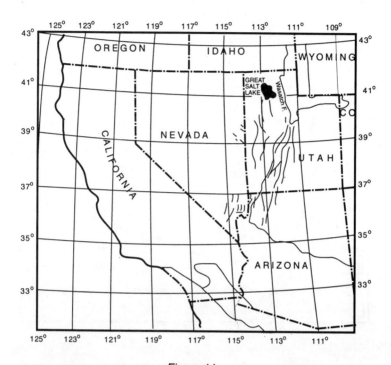

Figure 11
Partial map of the western United States, showing major fault zones in
Utah and Nevada only. (Continuous fine lines are streams.) Adapted
from footnote 17.

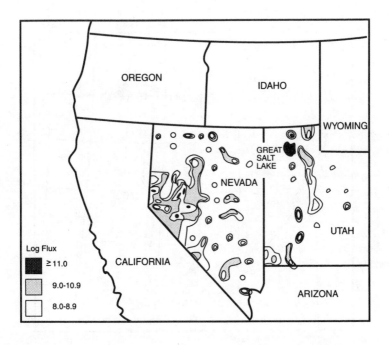

Figure 12
Map of estimated tectonic flux for Utah and Nevada, 1932 to 1966.
Adapted from footnote 17.

tating. The Yucca Mountain project is the U.S. Department of Energy's answer to a national problem. There it is charged with building a geologic repository (a large, deep underground mine) for the final disposal of commercial spent nuclear fuel and "high-level" radioactive waste. Current DOE plans call for keeping this waste secure and away from the public and the environment for 10,000 years, the time required for the most long-lived of the various radioactive isotopes to decay to safe levels. Most of the radioactive waste to be stored in the Yucca Mountain repository must be transported there by rail or truck from nuclear-electric power plants across the United States. Yucca Mountain is a small fault-block mountain composed entirely of volcanic rock called tuff. The repository is to be constructed in the relatively dry, unsaturated zone of the tuff, well above the water table.

This is a geohydrological setting that is expected to continue in effect for the required 10,000-year period, because the climate is semiarid and has been so for a long period geologically. But, even if this site were to survive earthquakes and breakup of the land, the shift in the poles would alter the climate, quite possibly to a more humid one that could cause the water table to rise and compromise the integrity of the repository.

If the land in the western states is "broken up" and if southern Nevada does suffer "an inundation by the earthquakes" near in time to the beginning of the significant Earth changes in 1998, then it would seem that the Yucca Mountain repository and the nuclear waste transportation scheme that DOE is developing will be trumped by natural events. This will leave a big problem whose solution may require developing a suitable waste repository in a safety land (see Chapter Eight).

It's possible to write a very large book on the seismotectonic potentials of all of the western states. Before we

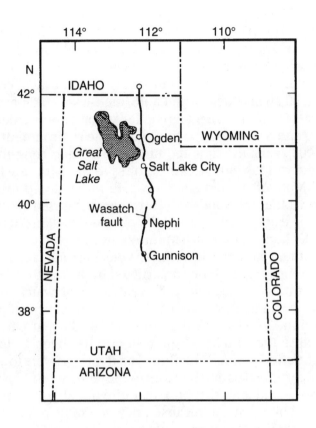

Figure 13
Map of 220-mile-long Wasatch fault zone and associated urban centers.

end our efforts, we'll just remind ourselves of the statement in reading 3976-15 that "*The Earth will be broken up in many places. The early portion will see a change in the physical aspect of the west coast of America*." In addition to the specific locations for potential breaking-up given in the first part of this section, recall that the Cayce readings often saw an acceleration in existing seismotectonic trends, rather than some totally new style or direction of Earth change. Thus, people living in the western states can take this into account when considering the potential for Earth-change destruction in the places where they live. Take historical seismotectonic and volcanologic information into account when assessing potential destructive effects. Specific information for one's location can be obtained from the U.S. Geological Survey or relevant state agencies.

THE CENTRAL AND EASTERN STATES

Two of the following readings indicate that disturbances in the central and eastern states will *follow* those occurring on the west coast:

> For, many portions of the east coast will be disturbed, as well as many portions of the west coast, as well as the central portion of the U. S.
>
> 1152-11
> August 13, 1941

> The Earth will be broken up in many places. The early portion will see a change in the physical aspect of the west coast of America.
>
> 3976-15
> January 19, 1934

> A Los Angeles, San Francisco, most all of these
> will be destroyed, before New York even.
>
> 1152-11
> August 13, 1941

It is difficult to convince people that there really *is* a high seismic risk in the central and eastern U.S. because there hasn't been a major earthquake there since the 1886 Charleston quake. While there have been several more-recent quakes whose magnitudes were large enough to cause loss of life and substantial property damage, fortunately their epicenters were in lightly populated areas. Of the three largest earthquakes in the central states in the twentieth century, two were on the Wabash Valley fault, which runs roughly north-south from the Ohio River along the Illinois-Indiana state line (Fig. 14). The quakes occurred in 1968 and 1987, and were of magnitude 5.5 and 5.2, respectively. A third, a 5.2 magnitude quake, occurred in 1980 near Sharpsburg, Kentucky. The largest quake this century to originate in the infamous New Madrid Seismic Zone was a magnitude 5.0 quake near Marked Tree, Arkansas, in 1976 (Fig. 14).

The New Madrid Seismic Zone

There are at least eight active known earthquake source regions in the Midwest, but only the New Madrid Seismic Zone has demonstrated a capability for strong-magnitude 6.0 or greater quakes during the past 300-400 years.[19] Most scientists and earthquake engineers consider the New Madrid Seismic Zone (NMSZ) to be a diffuse region starting from near Marked Tree, Arkansas, on the south to Cario and Metropolis, Illinois, on the north, an area of 120-150 miles trending northeast and 30-50

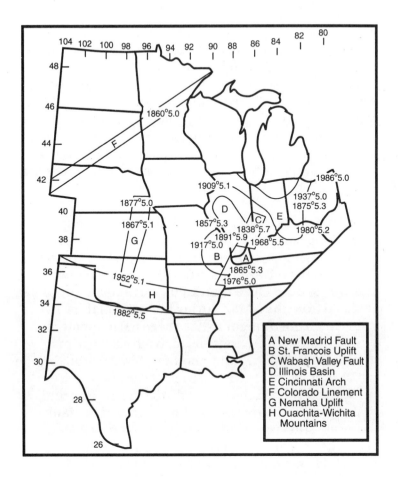

Figure 14
Earthquake source zones in the central United States, and epicenters
(small circles) and years of occurrence of all earthquakes between 1838
and 1986 with magnitudes from 5.0 to 5.9. New Madrid Seismic Zone
events of 1811-1812, 1843, and 1895 are excluded.

miles wide. This seismic zone is composed of the primary New Madrid fault and numerous secondary faults. The faults are for the most part buried beneath miles of sediments and sedimentary rocks and can be delineated only by geophysical mapping techniques. The NMSZ is known by seismologists and geologists, however, by the thousands of earthquakes that it has produced in recent centuries and by the thousands of square miles of ground surface that show the devastation of past great earthquakes.

> That the NMSZ is capable of major earthquakes and widespread, massive damage was decisively proven almost 200 years ago during a five-month period, from December 16, 1811, to the following May 1812 . . . More than 2,000 earthquakes originated from the NMSZ during that span of time, five of which are thought to have been 8.0 or greater in magnitude, 15 are considered to have been of magnitudes 6.7 to 7.7, and another 35 were probably 5.9 in magnitude or more . . . Thousands of scars scattered throughout the NMSZ from fissuring, liquefaction, lateral spreading, and surface faulting are still visible today, nearly two centuries later . . . They speak mute, but eloquent testimony of the cataclysmic upheavals in that time.[19]

The present inactivity of the other seven active earthquake regions in the Midwest doesn't necessarily mean that they could not become active again, for the Cayce readings predict that only selected portions of the Midwest will be spared disturbances:

> . . . the area where the entity is now located [Virginia Beach] will be among the safety lands, as will

be portions of what is now Ohio, Indiana and Illinois, and much of the southern portion of Canada and the eastern portion of Canada; while the western lands, much of that is to be disturbed—in this land [America]—as, of course, much in other lands.

1152-11
August 13, 1941

If we wish to speculate on the locations of safety lands in the Midwest, we might assume that the worst-case scenario for disturbances there will come from something like a repeat of the 1811-1812 devastation generated by great quakes in the NMSZ. A group[19] at Southeast Missouri State University has developed a CUSEIS (pronounced "Q-Seiz") or "Central United States Earthquake Intensity Scale." This intensity scale is used to assist local planners in the mitigation of earthquake damage from seismic waves from future large-to-great quakes generated within the NMSZ. The CUSEIS is specific to earthquake motions as they are transmitted through the rock types specific to the Midwest. Seismic wave energy in the frequency band of damaging ground motion (about 0.1 to 20 cycles per second) is reduced far less with distance traveled in the eastern states than in the western states. This smaller attenuation of seismic wave energy in the east is due to differences in rock absorptive processes (including also energy scattering) in the upper crust to a depth of about 12 miles. In the east, then, a magnitude 8 quake can produce strong Earth motions at relatively long distances from an epicenter in the NMSZ—or for one *elsewhere* east of the Rockies as well. For a hypothetical magnitude 7.2 quake, the areas of peak horizontal ground acceleration in the East are estimated to be up to ten times as large as those in the

West, and areas of peak ground velocity are five to ten times as large. This puts a large number of metropolitan areas of the eastern U.S. at greater risk than might ordinarily be considered.

The CUSEIS discussed above has been applied[19] to a number of maps for midwestern states. These maps give county-by-county estimates of maximum intensities to be expected from various magnitudes of quakes originating in the New Madrid fault system. One such map (without counties), for a worst-case scenario, is given in Figure 15. Note that *northern* Ohio, Indiana, and Illinois qualify as relatively safe under this scenario of midwestern disturbance. We'll digress for a moment to explain how one reads Figure 15.

The well-known Richter scale estimates earthquake *magnitude* or the amount of energy released at the underground source of the earthquake as measured by instruments like seismographs. The Modified Mercalli scale estimates earthquake *intensity* by assessing the *effects* of the quake at the ground surface. Such effects are determined by field surveys or by descriptions of damage obtained by sending out returnable post cards to people in earthquake-affected areas. There is no unvarying relationship between earthquake magnitude and intensity scales; however, the authors of the CUSEIS report[19] have developed a general relationship between the Richter and Mercalli scales for the specific types of quakes that occur in the NMSZ and for corresponding quake effects particular to the central and eastern U.S. Figure 15 is based upon that relationship. See Table 1 for a description of the Mercalli-scale earthquake effects for each of intensities VII through XII shown on the map.[19]

We must now consider the possibility that earthquakes like those of 1811-1812 could occur *elsewhere* in the eastern states. Sometimes large shocks occur in ar-

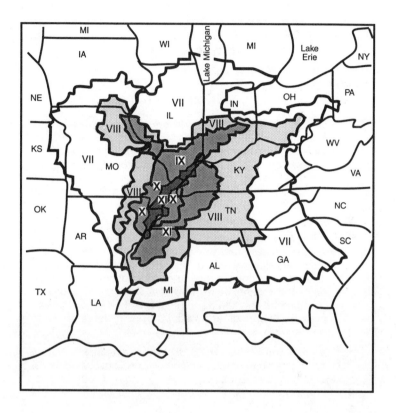

Figure 15
Great-earthquake-motion intensity map for earthquakes of Richter magnitudes 8.0 to 8.9 originating in the New Madrid Seismic Zone. Roman numerals refer to the Mercalli Intensity Scale of earthquake effects (Table 1). (Adapted from the map on page 59 of footnote 19.)

TABLE 1. MERCALLI EARTHQUAKE INTENSITY SCALE: GRADES VII—XII

Intensity
Value   Description

VII. Difficult to stand. Noticed by drivers. Hanging objects quiver. Furniture broken. Damage to masonry D, including cracks. Weak chimneys broken at roof line. Fall of plaster, loose bricks, stones, tiles, cornices, also unbraced parapets, and architectural ornaments. Some cracks in masonry C. Waves on ponds, water turbid with mud. Small slides and caving in along sand or gravel banks. Large bells ring.

VIII. Steering of cars affected. Damage to masonry C, partial collapse. Some damage to masonry B; none to masonry A. Fall of stucco and some masonry walls. Twisting, fall of chimneys, factory stacks, monuments, towers, elevated tanks. Frame houses moved on foundations if not bolted down; loose panel walls thrown out. Decayed piling broken off. Branches broken from trees. Changes in flow or temperature of springs and wells. Cracks in wet ground and on steep slopes.

IX.   General panic. Masonry D destroyed; masonry C heavily damaged, sometimes with complete collapse; masonry B seriously damaged. General damage to foundations. Frame structures, if not bolted, shifted off foundations. Frames racked. Serious damage to reservoirs. Underground pipes broken. Conspicuous cracks in ground. In alluviated areas, sand and mud ejected, earthquake fountains, sand craters.

X.   Most masonry and frame structures destroyed, along with their foundations. Some well-built wooden structures and bridges destroyed. Serious damage to dams, dikes, embankments. Large landslides. Water thrown on banks of canals, rivers, lakes, etc. Sand and mud shifted horizontally on beaches and flat land. Rails bent slightly.

XI.   Rails bent greatly. Underground pipelines completely out of order.

XII. Damage nearly total. Large rock masses displaced. Lines of sight and level distorted. Objects thrown into the air.

**TABLE 1 CONTINUED.
FOUR TYPES
OF MASONRY**

Masonry A. Good workmanship, mortar, and design. Reinforced laterally, and bound together using steel, concrete, etc.; designed to resist lateral forces.

Masonry B. Good workmanship and mortar; reinforced, but not designed in detail to resist lateral forces.

Masonry C. Ordinary workmanship and mortar; no extreme weaknesses like failing to tie in at corners, but not reinforced against horizontal forces.

Masonry D. Weak materials, such as adobe; poor mortar; low standards of workmanship; weak horizontally.

eas where quakes are relatively rare, as at Charleston, South Carolina, in 1886, or near the Grand Banks in the North Atlantic in 1929. Our history of earthquakes east of the Rockies goes back only about 100 years in the Plains states and about 300 years in the Atlantic coastal states. If we look for regions east of the Rockies that have been seismically quiet since the time of European settlement, a possible candidate could be the Meers fault, in southwestern Oklahoma, near Lawton. It shows evidence of ten feet of vertical offset that occurred about 1500 years ago. Such seismic quiet following a great quake can be seen in China where the great 1556 quake in Shanshi Province killed 830,000 people, a sizable portion of the world's population at that time in a region that at present is seismically very quiet.

Assuming that the historically recent past is the best indicator of seismicity in the near future we would place, in decreasing order, the regions of greatest earthquake threat in the eastern and central states as the NMSZ; the St. Lawrence Valley Seismic Zone (New York-Canada); the Charleston, South Carolina, area; the Cape Ann, Massachusetts, area; and the Wabash Valley Fault Zone of southwestern Indiana and southeastern Illinois. Recent discovery of an epicenter of at least one huge quake that happened 2,500 to 7,500 years ago near Vincennes, Indiana, prompted the statement, "Although the Wabash region has had only minor quakes in the last 200 years, it is seismically active. Southern and central Indiana should be considered vulnerable to severe shaking from an earthquake in the Wabash Valley."[20] At the 1996 meeting of the Seismological Society of America, geophysicists described a new linear structure running underground from near Little Rock, Arkansas, to Vincennes, Indiana. They said that the lineament could be a major fault zone, and the discovery of evidence of recent quakes at English Hills, Missouri, where a fault lies above the lineament, suggests that the lineament could be a major quake producer. "The lineament is an exceptionally long feature, and anything that is long in earthquake country is scary, because the longer a fault is, the greater the magnitude earthquakes that could occur on it." (*Science News*, April 6, 1996, p. 213)

Thus, the history of seismology is full of surprises insofar as the locations of damaging earthquakes are concerned. As reading 416-7 stated, times and places of future disturbances are "*as given by Him, 'As to the day and the hour, who knoweth? No one, save the Creative Forces.'*" What we do know, however, is that if the 1811-1812 quake scenario in the NMSZ were to repeat anytime soon, Memphis and St. Louis would be devastated and utilities, bridges, roads, and communications to the eastern states would be severed. Damage to buildings in St. Louis is projected,[21] for only a 7.6 magnitude NMSZ event within 100 miles of the city, at $2.9 billion. The quake would kill about 260 and injure 1,000. Many more billions of dollars damage could occur in Missouri, southern Illinois, Arkansas, Mississippi, Kentucky, and Tennessee.

The North Atlantic Seaboard

The following readings are relevant to the North Atlantic seaboard.

Q Will there be any physical changes in the Earth's surface in North America? If so, what sections will be affected, and how?
A All over the country we will find many physical changes of a minor or greater degree. The greater change, as we will find, in America, will be the North Atlantic Seaboard. Watch New York! Connecticut, and the like.
Q When will this be?
A In this period. As to just when—

311-8
April 9, 1932

Portions of the now east coast of New York, or New York City itself, will in the main disappear. This will be another generation, though, here; while the southern portions of Carolina, Georgia— these will disappear. This will be much sooner.
Q I have for many months felt that I should move away from New York City.
A This is well, as indicated. There is too much

unrest; there will continue to be the character of vibrations that to the body will be disturbing, and eventually those destructive forces there—though these will be in the next generation.

Q Will Los Angeles be safe?

A Los Angeles, San Francisco, most all of these will be among those that will be destroyed before New York even.

1152-11
August 13, 1941

What is there about the North Atlantic seaboard in general, and New York City in particular, that could make them vulnerable to Earth changes from a geoscientist's point of view? By analyzing old surveying records, a group of Stanford and USGS scientists has recently found evidence suggesting that during the last 100 years stress has been concentrating in the upper crust in western Long Island and just north of New York City. "Such stress accumulation, if maintained for several hundred years and then released suddenly, would generate a major earthquake."[22] A spokesman for the group concluded that "we just don't have any constraints on the timing or size of a future earthquake," if one occurs at all. The Hudson River area does lie at the northern end of a zone of seismic activity and larger earthquakes have struck there before. There was a shock in 1884 that was probably located at the mouth of the Hudson River and was felt from Maryland to New Hampshire.[22]

The moderator of an ABC-TV program ("Earthquakes: The Terrible Truth") that aired in 1994 said that the 1884 shock was estimated to have been magnitude 5 and that such a shock could be expected roughly every 100 years. He said that the fault that generated the 1884 quake cuts through Manhattan, Central Park, and beneath Greenwich Village. Columbia University seismologist Dr. Klaus Jacob said that a damaging quake in New York City is inevitable and that the city is not prepared to respond to it.

The Ramapo Fault Zone (RFZ) strikes northeast from Peapack, New Jersey, for about 60 miles into and across the Hudson River Highlands gorge. The RFZ lies adjacent to New York City. In a study involving radiocarbon dating of basal peats underlying tidal marshes, scientists found indications of both the sinking and rotating of crustal blocks within the RFZ in "a complex graben-like manner within the past few thousand years. "They also found that "the RFZ is an active fault zone with a recurrence interval of from a few hundred to 2,000 years . . . [and] . . . indeed, our data suggest that the RFZ will sustain a damaging movement within the next few decades."[25] If a "damaging movement" were significant enough to involve sinking and rotating of the crustal blocks beneath New York City there could be a quick fulfillment of reading 1152-11's prediction that " . . . New York City itself, will in the main disappear."

Another fault, the New York Bight fault lies beneath the ocean to east of New York City and is some 25 miles to it at its closest point. It may extend north beneath Long Island. Past motion on this near vertically dipping fault plane is downward to the west,[24] suggesting that the crust to the west under New York City may someday be subjected to downward motion.

As for such trends that might affect "Connecticut, and the like," relative changes in level-line data across the Connecticut valley suggest subsidence there.[25] There is a roughly 450-year record of seismicity for New England, and quake data show that coastal Connecticut and Massachusetts have experienced a number of significant events that have been large enough to cause damage. The quakes that occurred off Cape Ann, Massachusetts, on November 10, 1727, and November 18, 1755, have caused seismologists to classify eastern Massachusetts and Connecticut as an area of moderate seismic hazard with the potential for damaging shocks. Based on a continuing history of seismic activity there and the fact that most of the building stock was constructed before the adoption of seismic provisions to the building codes, there is ample reason for concern that the Boston area, for one, has a serious risk of damage, injury, and disruption should large quakes from the Cape Ann source strike again. Finally, in a study of crustal downwarping in coastal Maine, geologists have concluded that moderate earthquakes have occurred throughout historic time along the continental margin of northern New England but that quakes in Maine since 1975 have occurred in several clusters and in a zone that parallels the coast. Furthermore, "tide-gage and releveling data indicate that eastern Maine is downwarping at a greater rate than any other east coast locality. The increase in the rate of downwarping since 1940 may be related to a recent increase in seismicity in New England."[26]

A broad look at subsidence in the belt of Atlantic coastal states shows that areas of subsidence coincide with embayments along the ancient shoreline that existed in Late Cretaceous and Early Tertiary times.[27] These embayments or deep indentations of the ancient shoreline coincide with broad, concave-upward folds in the rocks (synclines) whose axes run transverse to the ancient shoreline. (See Fig. 16.) Embayment axes trend northwest, and the embayments have a history of subsidence since mid-Cretaceous times. The Raritan embayment is named after Raritan Bay, which lies off the south end of Staten Island, New York. "The subsiding Raritan embayment is the site of the largest earthquake in southeastern New York . . . and earthquakes occur around it

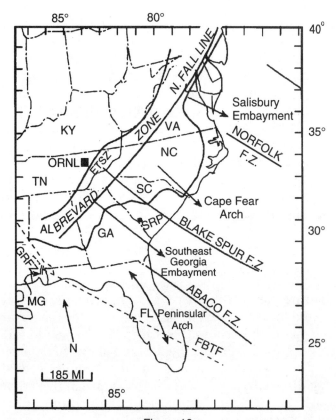

Figure 16

Map of the southeastern United States showing axes of embayments (see text) in the ancient shoreline of the North American continent. Major areas of seismicity are related to these gradually subsiding embayments; minor activity occurs on intervening arches. Trends of major fracture zones mapped on the Atlantic Ocean floor are shown where they are interpreted to exist on shore. The Brevard zone involves strike-slip faulting paralleling the Appalachian trend that becomes the Northern Fall Line fault zone. Explanations: ETSZ = Eastern Tennessee Seismic Zone; ORNL = Oak Ridge National Laboratory; SRP = Savannah River Plant; F.Z. = fracture zone; MG = Mobile Graben; GRFZ = Gulf Rim Fault Zone; FBTF = Florida-Bahamas Transfer Fault (interpreted); solid line near coast = generalized contact between Cretaceous and Tertiary coastal plain deposits; solid line interior = western side Valley and Ridge province. (Adapted from Fig. 10 of footnote 27.)

from northern New Jersey to southwest Connecticut. During the time of the larger earthquakes last century Raritan Bay had an unusual subsidence rate . . . "[27]

Thus, the ingredients for the natural destruction and disappearance of a major portion of New York City by earthquake and subsidence, respectively, seem to be in place. Here again, we are struck by the ability of "the sleeping prophet" to describe general geophysical conditions well before their discovery by scientists. And we see again that Edgar Cayce's Earth-change readings predict an acceleration in existing geological trends for portions of Connecticut and New York, including New York City.

Eastern Tennessee

During the last decade more seismic energy has been released (per unit area) in eastern Tennessee than in any other part of the eastern U.S., except for the New Madrid, Missouri, area. Most of the seismic activity is concentrated along a line between Chattanooga and Knoxville, coinciding with a major boundary between two crustal blocks. This Eastern Tennessee Seismic Zone (ETSZ) could represent a major fault capable of producing large earthquakes (*Science News,* v. 145, p. 278, 4/30/94). The ETSZ (see Fig. 16) runs near several hydroelectric dams, two nuclear power plants, and the Oak Ridge National Laboratory (ORNL, Fig. 16), located about 22 miles west of Knoxville.

South Carolina and Georgia

As of this writing, there is no evidence for significant subsidence of southern coastal South Carolina and Georgia, despite this Cayce-reading forecast:

> **Portions of the now east coast of New York, or New York City itself, will in the main disappear. This will be another generation, though, here; while the southern portions of Carolina, Georgia—these will disappear. This will be much sooner.**
>
> **1152-11**
> **August 13, 1941**

Only minor sinking or loss of coastal land has been reported for South Carolina. The U.S. Army Corps of Engineers summarized shoreline movements for the stretch of coast between Tybee Island, Georgia, and Cape Fear, North Carolina, for the period 1851-1983 (Tech. Rept. CERC-83-1). It showed that more than 50 percent of the shoreline between Tybee Island, lying on the Georgia-South Carolina boundary, and St. Helena Sound, lying due east of Beaufort, South Carolina, had been eroding at greater than one meter per year. This erosion rate is the greatest for the overall stretch of coast studied, and although Corps authors suggest that the erosion is due to wave interactions with a mostly north-south-oriented coastline, it may also reflect slow coastal subsidence there. I could find no measurement-based reports of coastal subsidence or erosion for Georgia, but G.H. Davis, and others, determined that the Savannah area had subsided about four inches over the period 1918-1955, due apparently to land subsidence from pumping and subsequent decline in artesian head in the underlying Ocala limestone (*Bull. Geological Society of America,* 1959, v. 70, p. 1585).

What might have Cayce been seeing, in reading 1152-11, as the fundamental tectonic reason behind the predictions of subsidence of the southeast coast from about Charleston southward to the Georgia-Florida line? Figure 16 shows that this stretch of coast is bounded by projections of two northwest-trending fracture zones that occur on the floor of the western Atlantic Ocean. To the north, the Blake Spur fracture zone is believed to coincide with the Charleston seismic zone, while to the south, a northwest-trending fault zone beneath the coastal plain sediments of southeast Georgia "coincides with the landward projection of the Abaco [also called the Jacksonville] fracture zone.[28] The Southeast Georgia embayment appears to lie over a crustal block dropped between the two major fracture zones. Coastal seismicity is concentrated at the flanks of the Southeast Georgia embayment, which is "relatively *subsiding* at present"[27] [emphasis added].

And so here we have (1) bounding, northwest-trending fracture zones and (2) a gradually subsiding embayment, as the tectonic prerequisites for submergence of "the southern portions of Carolina, Georgia" as described in reading 1152-11. And if mantle movements associated with an oncoming pole shift begin to raise the sea floor in the Atlantic, as per reading 3976-15, off the U.S. east coast (see below), one might well expect an acceleration of some of the existing long-term trends in coastal subsidence along the present Atlantic Seaboard. This could be especially true for such ancient synclines as the Raritan and Southeast Georgia embayments.

SOCIETAL TRENDS AND AMERICA'S FUTURE

Cayce was not the only prophet of dire events in America's future. Prophecies of Native American medicine men and international clairvoyants, from the French monastic Nostradamus to the Tibetan mystic Phylos, suggest that America faces stormy times in the near future. Of these, the forecasts of Phylos, spelled out in a book, *A Dweller on Two Planets,*[29] published in 1886, seem to parallel Cayce's most closely.

Phylos, a disincarnate spirit who dictated his story to the author of the book, Frederick Oliver, claimed to have had many incarnations, in Tibet, on Atlantis, and most recently in America during the California gold rush. Between Earthly incarnations, his soul spent time in the realm of Venus, thus his claim of having dwelled on two planets.

Phylos's story is relevant for several reasons, one being his descriptions of the technological accomplishments of the Atlantean civilization. Conveniences on Atlantis are said to have included transportation by antigravity devices, "telephotic" service, similar to our television, and a "caloriveyant" instrument by which energy could be transmitted by Earth currents and converted at distance into heat. In the words of Phylos, this energy transmission also utilized "those [characteristics] of the higher ether, a range which ye shall yet find and utilize as did Atl [Atlantis], for are ye not [Atlantis] returned? I have said it. Ye lived then, ye live now. Ye used all these forces then; ye shall ere long use them all again."

Phylos makes many references to America as being "Atlantis come again," a theme that parallels a Cayce reading:

> Be it true that there is the fact of reincarnation, and that souls that once occupied such an environ [Atlantis] are entering the Earth's sphere and inhabiting individuals in the present, is it any wonder that—if they made such alterations in the affairs of the Earth in their day, as to bring destruction upon themselves—if they are entering now, they might make many changes in the affairs of peoples and individuals in the present?
>
> 364-1
> February 3, 1932

Earlier this Cayce reading said, "As we recognize, there has been considerable given respecting such a lost continent by those channels such as the writer of Two Planets . . . " that is, as given by Phylos.

What was the nature of Phylos's forecast? He begins his ruminations on the end of the present age with words full of karmic penalty:

> **Atl, just because she is Atl returned on a higher plane, she must endure the woes as well as retrace her precarnate glories. The penalty visited upon Poseid [the last remaining large island of Atlantis] was the crowning sentence of that Age. Century after century in the majestic march of Time hath passed since the sun looked down upon a wild waste of ocean waters where but a few days before had been the regal Island-Continent. Another cycle hath reached its end, and its last now-closed Sixth Day is come, in stately, measured, but inexorable way to face judgment by the standard, Truth.**

Phylos then asserts that there is no escape from the "awful woe" of the end of the age, as pictured in prophecy over many centuries. He asserts that "America, the Glorious, together with the rest of the world" will meet worse woe than Atlantis, "though not by water but by fire." He goes on, "The Hour hath struck. And yet in all of this there is no mystery, no supernatural penalty, no capricious infliction by an offended personal God, and nothing of 'man's necessity, God's opportunity.' It is all of Man's own doing. He hath wandered from the Way, and hath for the God-nature in him, which he should have revered and nourished, substituted worship of Self and of Mammon; hath cast out Love; and placed violence,

lust, greed and all the riotous animalism in him in command of his life. *Man is his own judge and executioner.*"
Phylos then proceeds to describe the end of our age:

Soon millions of trained soldiers will turn upon the visible representatives, the wealthy and worldly prosperous, who in reality are not more responsible than will be their assailants, of that Relentless Force [selfish action] behind all human enterprise. Later they will break up into lawless bands bent on satisfying Ishmaelitish tendencies, each self-server's hand weaponed against their fellow creatures. Then will the pent-up hate, the savagery and selfishness begotten by ages of selfishness ruled by unbridled animalism break in a storm such as the world hath never yet seen ...

In what appears to be a singularly prescient statement (see italics below) he continues:

Friends, know ye the meaning of the name Jerusalem? That it meanth "Vision of Peace"? Verily, so it doth. One by one during the years all the signs of the end of the Age but one were fulfilled; but these were "only the beginning of sorrows," for still the Spirit of Liberty abode here and there in the breasts of lovers of their fellowmen. The Spirit wrapped itself in the glorious folds of the Stars and Stripes and proclaimed the imperishable declaration of human equality, granting unto all that freedom which Americans for themselves demanded. *But now the "Vision of Peace" is finally encompassed by armies, the last gap being filled with blue-coated soldiers forcing Mammon's commercial shackles upon alien peoples in tropical islands ...* **"Then shall the end come." A Son hath continually called from**

on High: "Stand from under! Get into the shelter of that Cross."

For today's world situation, could it be that "blue-coated soldiers" refers to businessmen, and perhaps bureaucrats and/or local police, and that the "alien peoples in tropical islands" refers primarily to the inhabitants of Southeast Asia. The "commercial shackles" forced upon these peoples could be the servitude, often prisonlike working conditions, and minuscule wages experienced by laborers in Malaysia, Hong Kong, Singapore, and Indonesia. These conditions have been documented by the American press.[30] One article, entitled "Asian Labor: Wages of Shame," is headlined, "Western Firms Help to Exploit Brutal Conditions." Children are often the only wage earners in some areas of Southeast Asia because adult workers have been laid off in favor of children, who are infinitely more exploitable and provide larger profits for wealthy factory owners and corporate stockholders worldwide. Closer to home, on the tropical island of Haiti, we have a situation in which, "While the Clinton administration is willing to risk the lives of 20,000 American soldiers in Haiti, it is not willing to risk the profits of American textile manufacturers."[31] (On Oct. 21, 1994, the U.S. reimposed quotas on imports of Haitian clothing. This seriously harms one of Haiti's most important industries and one of its largest earners of U.S. dollars.)

As to Phylos's references to violence, lust, and animalism in American society, start with the violence and eroticism displayed routinely on American TV and now exported to ever larger audiences worldwide. Then open the *Wall Street Journal* and read about "a publicly held firm that turns X-rated videos into a hot business"[32] or about "porn lines [that] offer 800 numbers."[33] Now consider that the U.S. military-industrial complex makes more money on weapons sales overseas than it does on domestic sales, and some of these sales are financed by U.S. Government loans. By 1980, the top ten U. S. corporate earners involved in foreign weapons sales were making $15 billion yearly, and they ranged from General Dynamics to Chrysler Corporation.[34] Consider that in 1995, the U.S. controlled 53 percent of the world arms trade and has been called "the world's leading merchant of death."[35] American exports of antipersonnel mines have maimed and killed thousands of children over the last few years. America also leads the world in per capita incarceration of its own citizens.

And what of Phylos's vision of "trained soldiers (turning) upon the visible representatives" of society and of lawless bands forming to satisfy selfish goals? An *Atlantic Monthly* article,[36] entitled "The Coming Anarchy," refers to just such activity in Sierra Leone. It suggests that "West Africa is becoming *the* symbol of worldwide demographic, environmental, and societal stress, in which criminal anarchy emerges as the real 'strategic' danger. Disease, overpopulation, unprovoked crime, scarcity of resources, refugee migrations, the increasing erosion of nation-states and international borders, and *the empowerment of private armies* [emphasis added], security firms, and international drug cartels are now most tellingly demonstrated through a West African prism. In Sierra Leone, the government force fighting the rebels is full of renegade commanders who have aligned themselves with disaffected village chiefs. A pre-modern formlessness governs the battlefield, evoking the wars in medieval Europe . . . "

Is there any evidence in the United States of "trained soldiers (turning) upon the visible representatives" of society? Unfortunately, evidence of this tendency is growing. In addition to former U.S. soldiers joining private militias nationwide, consider the recent report[37] on an underground group claiming to be serving the army's elite Special Forces, headquartered at Ft. Bragg, N. C. The group has been publishing a far-right dissident newspaper, *The Resister,* that is extremely critical of the federal government. A *New York Times* article concludes that "extremists trained in special warfare are seen as posing a serious problem" to society. And we remember, too, that an army soldier was in custody in 1995, charged with blowing up the Federal Building in Oklahoma City.

When Cayce was asked whether Phylos's vision of the end of the age was "based on truth, especially in its forecast for the United States, the answer was:

> **As viewed by an entity separated from the whole, yes. As truth that may be implied by one that looks only to the Lamb, to the Son as a leader, no. Choose thou.**
>
>

282-5
March 2, 1933

It appears that the "entity" referred to here is Phylos, and that Phylos's understanding is correct, but limited, however slightly, by such clouded vision as can affect any soul not fully absorbed into the whole. I interpret this reading to mean that karmic penalty, as enunciated by Phylos, may be mitigated for a nation that does an about face, and starts to try to implement Christlike principles in its activities at home and abroad. Thus, the law of karma can be superseded by the law of grace, as intimated in this reading:

> **As in relationship to changes, these are indicated not only through prophecies but through astrological aspects, as well as the thought and intent**

of persons and groups in high places; bringing about these things, these conditions, in what might be said to be the fullness of time.

However—since the advent of the Son of Man in the Earth, giving man an advocate with the Father, there has been an influence that may counteract much of that which has been indicated that would come as retribution, or in filling the law of an evolution of ideas and the relationship of material things to the thoughts and intents of individuals and groups.

Then, as to whether the hearts and minds of individuals or souls (who were given authority concerning the laws of the universe) are fired with the thoughts of dire consequences or those things that bespeak of the greater development of a spiritual awakening, is still in the keeping and in the activities of individuals who—as this entity—have caught a glimpse, or an awareness, of that which is in the making, in the affairs of state, nation, and nations, and the universe, as related to the conditions upon the face of Mother Earth.

There enters much, then, that might become questions as respecting . . . the activities of groups and individuals *who have acted and who are to act* [emphasis added] as a counterbalance to these happenings in the Earth . . .

Again the interpretation of the signs and the omens becomes an individual experience. And each soul—as this entity—then is given the privilege, the opportunity to *live* such an activity in its relationships to its fellow man; filling, fulfilling, and interpreting that which has been indicated, in such measures and such manners as to bring hope and not fear, peace and not hate, that which is *con-*

structive not destructive, into the lives and minds and hearts of others.

1602-5
November 28, 1939

Before leaving the subject of what's ahead for America, the following reading forecasts:

> Ye are to have turmoils—ye are to have strife be-
> tween capital and labor. Ye are to have a division in
> thine own land before there is the second of the
> Presidents that next will not live through his of-
> fice—a mob rule!

3976-24
June 16, 1939

Has the predicted strife, division, and mob rule in America already occurred, or is it yet to come as part of the events surrounding the end of the age?

The answer seems most to depend upon which president the reading is referring to. The president at the time of the reading was Franklin D. Roosevelt. Inasmuch as the date of the reading is 1939, does the reading pinpoint a period *after* FDR's death and *before* John F. Kennedy's? Or do the words "that next" in the reading dictate a different interpretation, one which would require two presidents to die in office *after FDR?* We'll consider both possibilities, but emphasize the latter for reasons given later.

According to *Webster's, next* means "immediately preceding or following (as in place, rank, or time)." And so, what is the starting point in time from which one determines *next?* Are we talking about the second of the presidents after Garfield, that next will not live through office? That is, are we talking about the second of the presidents *including* Roosevelt or the second next *after* Roosevelt? If our starting point is the date of the reading, 1939, then the first of the presidents next to die in office was Roosevelt, and the second of the presidents was Kennedy. This chronology, however, runs afoul of the fact that the United States did not experience a significant *"division in thy own land"* between 1939 and the second of the president's (JFK's) death in office in 1963. Some might argue that the isolationists' attacks on Roosevelt for modifying the neutrality laws and for his trying to get the U.S. to enter World War II after Germany's invasion of Poland on September 1, 1939, might be the division referred to. Others might argue that the civil rights initiatives and related socio-political turmoil between 1939 and 1963 would qualify for the "division" seen. However, these are relatively mild by comparison to the Watts riots and the social consequences of the Martin Luther King assassination that took place *after* JFK's death in 1963. Indeed, it was the Kerner Commission that, *in 1968*, five years after JFK's death, reported that "our nation is moving toward two societies, one black, one white—separate and unequal." Compare that relatively bland pronouncement with today's headlines about America being two nations, separate and hostile, and not for just racial but for economic, ethnic, gender, and age reasons.

One would also have to find evidence of significant *"strife between capital and labor"* during the time between 1939 and JFK's death to get complete agreement with the reading's prognostications, if JFK were the "second of the Presidents." The labor relations record of President Truman's years provides some evidence for such strife. Serious strikes in several heavy industries led to a politico-economic crisis in August, 1946. Truman asked Congress for a law to draft into the armed forces anyone refusing to work in industries taken over by the government. Congress refused to grant the law, but Truman's request embittered certain labor leaders. Also in 1946, the government used an injunction to prevent a coal strike. When John L. Lewis, head of the United Mine Workers, ignored the order, he and the union were fined more than $3 million by a federal court. Finally, to avoid a strike, Truman seized the steel industry in April 1952. This precipitated a grave constitutional issue, settled only by the Supreme Court, which ruled the seizure unconstitutional. This record of "strife" between capital and labor seems to fit the reading's prediction, and it does fall into the critical 1939-1963 period. But what doesn't fit is the prediction of a "division" in our own land that seems yet to take place. Perhaps the only way to resolve the matter is within the context of the last three words of the reading . . . "a mob rule!"

There seems little question that the mob rule of the reading refers to the demise of the second of the presidents. In Kennedy's case, much has been speculated to the effect that his assassination was the result of disgruntled mobsters. A recent book entitled *Case Closed*,[38] agrees with the Warren Commission's conclusion that Lee Harvey Oswald was the lone assassin of Kennedy and makes a case for the lack of credible evidence that mobsters were behind the evil deed. The book's conclusions have been widely accepted, although some may still be tempted to argue otherwise. Nevertheless, it seems reasonable to assume that whatever criminality existed in Kennedy's time was not a sufficient source of mob rule to result in his death, but was merely emblematic of a trend that is increasing over time. This assumption accepted, we are free to speculate that the enigmatic word "next" refers to the death in office of one more president *after* Kennedy and that strife between capital and labor will pick up again before this happens.

What about future prospects for strife between capital and labor? The U.S. has had only a minor amount of labor strife since Kennedy's death (1963). According to the history of work stoppages since 1965,[39] there has been a slow drop in worker-days idle as a percent of estimated working time until, in 1992, it reached a low of 0.01 percent. However, just recently, there has been an increase in high-visibility, capital/labor strife such as the baseball- and hockey-player strikes, strikes in the coal industry, and at Caterpillar, Boeing, and General Motors. And what about the feelings of a U.S. labor force that's been pushed toward the upper limits of capacity without 1980s-style wage increases?

Record-length workweeks, record overtime hours, deferred vacations, plus three years of forced increases in productivity mean high stress. Morgan Stanley's chief economist, Stephen Roach, believes that the U.S. is coming off the bottom in the wage cycle. He notes that 1996 will see a number of major labor contracts expiring, including the United Auto Worker's pact with Detroit's Big Three:

> **This is a very heavy collective-bargaining year, with 32% of all union workers renegotiating their contracts. Workers, whether they are in unions or not, are feeling increasingly cut off from this so-called prosperity of the nineties. And they are going to make an effort to recoup what they perceive to be their fair share.**[40]

Note, too, that in 1995 John Sweeney was elected the new president of the AFL-CIO. Mr. Roach says that Sweeney is "probably the most militant labor leader we have had in the United States since the 1950s."[41]

In February 1998, thousands of Indonesian workers rioted due to sweeping austerity measures imposed under a $40 billion International Monetary Fund bailout. The bailout followed a huge plunge in the value of the rupiah. Strife between capital and labor is also expected in Korea and other southeast Asian countries affected by currency instabilities. Could not America be far behind?

Q Would it be feasible to work out an international currency, or . . . stabilization of exchange values?

A This, too, will be worked toward. It will be a long, long time before established. There may indeed be another war over just such conditions . . . (3976-28, 6/20/43); [and] when the present conditions [WW II] have subsided . . . there will be more and more upsetting in the monetary units in the land (5400-1, 8/21/44).[41]

To return to the subject, what kind of future *"division in thy own land"* could the reading be envisioning? A geological or hydrological division, such as a crustal split or enhanced waterway down the center of the U.S., seems highly improbable, despite the prognostications of certain psychics in recent years.[42] For example, in the book, *We Are the Earthquake Generation*,[43] a consensus of psychic predictions was used to draw a map of the U.S. for the 1990-2000 A.D. period. It shows portions of Wisconsin, Michigan, and Illinois under the waters of the Great Lakes. The Great Lakes and a much enlarged Mississippi River drainage way are described as "new ocean areas" that divide the United States into two separate land masses.

There seems to be little doubt that the inspiration for this hydrogeological division of the U.S. comes from the following Cayce reading:

The waters of the lakes [Great Lakes] will empty into the Gulf [Gulf of Mexico], rather than the waterway over which such discussions have been recently been made [St. Lawrence Seaway].

1152-11
August 13, 1941

But there is no reason to require the ocean to fill (1) the St. Lawrence Seaway, (2) the Great Lakes, (3) a new waterway along the Illinois-Indiana state line, and (4) the entire Mississippi waterway from the southern tip of Illinois to the Gulf for this reading to be realized. Consider rather the following quote from C.R. Longwell and others (1939, *A Textbook of Geology*, Wiley, N.Y.):

An excellent example of [crustal] tilting on a large scale is afforded by the Great Lakes. To the northeast the land has risen since the disappearance of the great ice sheet, and as a result the lake basins have been tilted southwestward. The tilting movement is still in progress and has been accurately determined; it is at the rate of five inches per hundred miles per century. Small as this rate seems, in 1600 years it would cause the upper Great Lakes [Superior, Huron, and Michigan] to discharge by way of the Chicago River into the Mississippi drainage.

And, as provided by C.E. Larsen in a further study of tilting of the Great Lakes area (1987, U.S. Geological Survey Bulletin 1801):

" ... [crustal] uplift finally raised the North Bay [Ontario] sill [then draining the upper lakes] above southern outlets between 4,500 and 4,000 yr. B.P.

The Chicago outlet was abandoned after 4,000 yr. B.P. and the modern drainage system ... [through the St. Marys, St. Clair, Detroit, Niagara, and St. Lawrence Rivers] came into being. [But water levels in] Lake Erie and Lake Ontario continue to rise in concert with their uplifting eastern outlets.

Thus, the prediction of the fragment from reading 1152-11 merely accelerates an ongoing geologic trend, and no subsidence of the crust and invasion of the ocean are required for the Great Lakes to empty into the Gulf of Mexico.

Assuming, then, that "a division in thy own land" does not refer to anything geophysical, we are free to speculate upon social situations that have the potential to further divide our sense of community and lead to extensive mob rule, as in uncontrollable mob violence like the Rodney King riots in Los Angeles or to mob rule, as in the corruption of politicians, bureaucrats, and law-enforcement institutions by gangs or organized crime.

In the former case, we have heard that massive immigration will destroy America,[44] and the overwhelming passage of the California referendum on immigration (Proposition 187) has set up a bitterly divisive social battle in that state. The division between those who are for and those who are against large-scale immigration is as contentious as is the one between those for and against abortion-on-demand. Then we have the consciously segregated districting system whose construction was recently attempted in the name of the Voting Rights Act. "Our drive to segregate political districts by race can only serve to deepen racial divisions by destroying any need for voters or candidates to build bridges between racial groups or to form voting coalitions" (Justice Clarence Thomas in *Holder vs. Hall,* a 1994 Supreme Court case).

To these examples of social divisions we can add many other national and local issues that divide us, from school choice to decriminalization of the use of illegal drugs. Perhaps most important, will the U.S. be further and irrevocably divided into a nation of moral haves and aggressive, morally bankrupt have-nots? Consider Republican presidential aspirant Pat Buchanan's view:

> **Where did that L.A. mob come from? It came out of public schools from which the Bible and the Ten Commandments were long ago expelled. It came out of drugstores where pornography is everywhere on the magazine rack. It came out of movie theaters and away from TV sets where sex and violence are romanticized. It came out of rock concerts where rap music extols raw lust and cop-killing. It came out of churches that long ago gave themselves up to social action, and it came out of families that never existed.**
>
> **When the Rodney King verdict came down and the rage boiled within, these young men had no answer within themselves to the questions. Why not? Why not loot and burn? Why not settle accounts with the Koreans. Why not lynch somebody—and get even for Rodney King?[45]**

In the latter case, of corruption of politicians and government employees, "mob rule" may be the result of the rapidly growing drug gangs, who compare favorably to the barbarian gangs that sacked the Roman Empire 15 centuries ago. Increasing drug-gang activity has been reported by author James Davidson:

> **The drug lords have incentives like almost no one else to become involved in the political pro-**

cess. And they have the ready cash to buy what they want. What they want is more corruption. They want to buy exemption from the law.

Unlike a legitimate business, the drug cartels operate in an almost entirely clandestine fashion. They must. Murder is part of their business. Where they cannot buy a police investigator or bribe a witness who has incriminating evidence, they frequently resort to violence.

... the real drug lords are not buying aldermen, they are buying governors and congressmen and the leaders of countries. Or shooting them.[46]

Thus, could drug-gang "mob rule" eventually be the reading's vision for the possible death in office of the second of the presidents after Roosevelt? The net profit of the cocaine trade alone is authoritatively estimated at about $100 billion per year and the United States is the largest market for cocaine. As Davidson points out, "It is also bound to be the place where a large part of the cocaine profit comes to rest. It is therefore likely to be in the United States where the impact of corruption is the greatest, even if that is not yet obvious. The sums are immense. They rival the surplus of Japan."[47]

Finally, a recent lead editorial in the *Wall Street Journal* is entitled "Investigate Mena." Mena is the tiny airfield in western Arkansas around which a clouded tale of drug smuggling and spy operations grew in the 1980s. "But the big story here is not primarily about who did what 10 years ago. It's about a very 1990s concern: drugs. How has our system broken down so that illegal drugs can be moved into this country on such a large scale?"[48]

We are left with the disquieting feeling that, just perhaps, the division in our own land is yet to come. And that it might have to do with corruption of the U.S. political, legal, and law-enforcement systems due to drug cartels (mob rule) and their increasingly dangerous manifestations in our society and the world at large. Consider this sampling of recent newspaper reports:

- **Trial of the century to probe mob grip on Italy's leaders.**[49]
- **Global gangs**[50]
- **Mexico drug lords gain more power—Nation could be next Colombia**[51]
- **Russian organized crime goes global**[52]
- **Scandal in Colombia emerging—Government linked to Cali drug cartel**[53]
- **Five nights a week, at least $100 million in crisp new $100 bills is flown from JFK nonstop to Moscow, where it is used to finance the Russian mob's vast and growing crime syndicate. State and federal officials believe it is part of a multi-billion-dollar money-laundering operation. The Republic National Bank and the United States Federal Reserve prefer not to think so.**[54]

Finally, we note that increasingly over the last five years, drug smuggling is giving way to illicit trade in uranium and plutonium, affecting international security. In a recent article[55] in *Scientific American*, two scientists discuss what they call "The Real Threat of Nuclear Smuggling." They include the following observations:

> **... there is a clear danger that organized crime groups or terrorists could also join the nuclear club. The transition from transporting nuclear contraband to using it directly is apparently an easy one: radioactive isotopes have already been used for murder. In late 1993, Russian "Mafia" as-**

sassins allegedly planted gamma-ray-emitting
pellets in the office of a Moscow businessman, killing him within months.

The authors go on to describe how criminal organiza-
tions could also use radionuclides for large-scale extor-
tion against a government or corporation.

And so, we are left with the question, "Is mob rule,
leading to the death in office of one of our presidents,
yet to come?" Or will further research demonstrate that
John F. Kennedy was indeed the second of the presidents, killed by mobsters,[56] and thus the "divisions" visualized by Cayce in 1939 have already occurred?

5

Will Japan Be Submerged?

Scientists calculate that about 15 percent of all the seismic energy released on the planet is focused on Japan, where 120 million people are crowded onto an archipelago with a total land area about the size of California's. Japan has been shaken by earthquakes for all of recorded history. Between 1603 and the great Kanto district (Tokyo) quake of 1923, about 40 damaging shocks occurred. Some of these were followed by tsunamis of various destructive power. Active volcanoes dot the Japanese archipelago as well, producing numerous eruptions.

The underlying mechanisms for these unwelcome events are now known to be related primarily to the convergence of the margins of four crustal plates, beneath or adjacent to Japan. The Eurasian plate lies to the west, the North American plate extends down from the north, the Pacific plate pushes in from the east, and the Philippine Sea plate shoves northward at the geologically rapid rate of 1.5 inches per year. (See Fig. 17.)

The prediction that "the greater portion of Japan *must* go into the sea" [italics added] is found in the 1934 reading, communicated by Halaliel, that describes many other large-scale Earth changes as well:

> **As to the changes physical again: The Earth will be broken up in the western portion of America. The greater portion of Japan must go into the sea. The upper portion of Europe will be changed as in the twinkling of an eye. Land will appear off the east coast of America. There will be the upheavals in the Arctic and in the Antarctic that will make for the eruption of volcanoes in the Torrid areas, and there will be shifting then of the poles ...**
>
> **3976-15**
> **January 19, 1934**

The major geophysical problem for Japan is subduction, a process in which one lithospheric plate descends beneath another. We must make a distinction, however, between gradual descent of one or more lithospheric plates beneath Japan and the readings-implied acceleration of this process that could result in rapid subduction and submergence of much of northern Honshu and Hokkaido Islands, or "the greater portion of Japan." Destructive earthquakes, tsunamis, and volcanic eruptions may occur in Japan between now and the end of the century, even without acceleration of the subduction process, but they will reach unprecedented proportions if the process does accelerate.

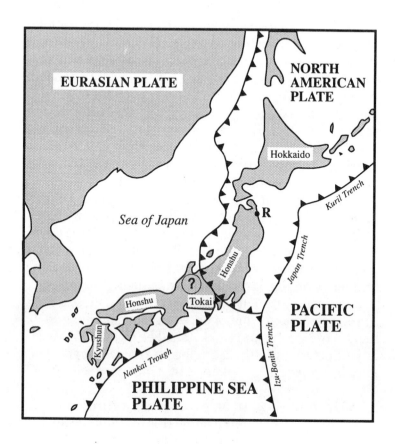

Figure 17
Crustal plates and their motions in the general geodynamic setting of
Japan. R = Rokkasho. The question mark denotes uncertainty in the
location of the boundary between the North American and Eurasian
plates in the mountains of central Japan. Solid triangles point in the
direction of plate movement.

The Pacific plate has, for a relatively long time geologically, been subducting to the west beneath northern Honshu and Hokkaido Islands. A "Benioff fault plane" has developed that extends to a 435-mile depth beneath the Sea of Japan, on a 45° dip from where it originates beneath the Japan Trench (Fig. 17). According to the theory of plate tectonics and sea-floor spreading, the Pacific plate is sinking into the mantle beneath Japan, causing earthquakes and erupting volcanoes along the upper boundary of the plate.

Earthquakes such as the 1978 Miyagi-Oki (M 7.6) and 1986 Tokachi-Oki (M 8.2) events are examples of quakes due to the Pacific plate underthrusting Japan. If the Cayce Earth-change readings are any guide, as 1998 approaches great underthrusting quakes will occur with increasing frequency on the east coast of Honshu and Hokkaido.

But the important tectonic story in Japan these days is a newly recognized kind of underthrusting on the *western* margin of Honshu and Hokkaido. (See Fig. 18.) This easterly directed underthrusting has produced large, *shallow* earthquakes, as explained by a Japanese scientist:

> The August 2, 1940, earthquake ruptured the northernmost segment along this margin; the June 16, 1964, Niigata earthquake ruptured the southernmost segment; and the May 26, 1983, Japan Sea earthquake then filled in a segment off northern Honshu . . . It now appears that the July 12, 1993, event ruptured the segment between the 1940 and 1983 earthquakes.
>
> Careful study of the focal mechanisms of the 1940, 1964, and 1983 earthquakes are all consistent with the Sea of Japan [Eurasian plate] thrusting beneath Honshu and Hokkaido. Since there is no deep [Benioff zone] associated with the under-

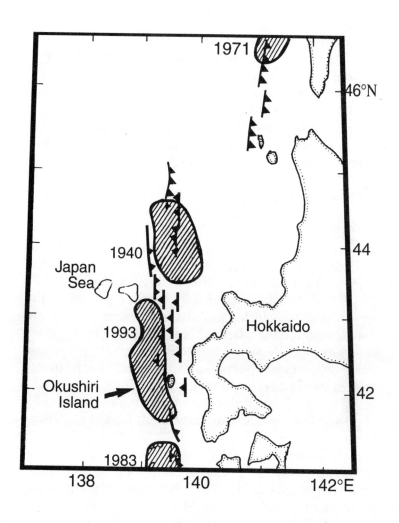

Figure 18
Source areas of large earthquakes (shaded areas) near Hokkaido since 1940, and distribution of active thrust faults. (Simplified and redrawn from footnote 43.)

**thrusting of the Sea of Japan, one simple interpre-
tation of the tectonics is that a new subduction
zone is forming off the west coast of Honshu and
Hokkaido . . . Thus, a closer look at this region
shows that the July 12, 1993, event [the Hokkaido-
Nansei-Oki quake] is not on the "wrong side" of Japan—instead, Japan is unfairly burdened with
active underthrusting on both sides![1]**

Thus, the subduction off the west coast of Honshu and
Hokkaido Islands seems to have started within the recent geologic past, on a fault plane that is dipping to the
east under Japan. This fault plane dips under the part of
the North American plate that extends down to central
Honshu from the north. (See Fig. 17.) The July 12, 1993,
magnitude 7.8 quake on this fault produced one of the
largest tsunamis in Japan's history, a seismic sea wave
that devastated the island of Okushiri (Fig. 18) in the Sea
of Japan. If the Cayce Earth-change readings are any
guide, Japan can expect an increasing number of severe
quakes due to acceleration of subducting crustal plates
along both its east and west coasts. One recent *Pacific*-
plate subducting event took place off Japan's northeast
coast on October 4, 1994.[2]

If the newly recognized, shallow-subducting mechanism caused by the Eurasian plate moving downward
beneath the edge of the North American plate were the
principal subduction process to accelerate, then one could
speculate that much of northern Japan, on up to and including Sakalin, could begin moving along the newly
developing fault zone into the Pacific. On the other hand,
acceleration of the subduction of northern Honshu and
Hokkaido might be accomplished mainly by sudden acceleration in the westward movement of the Pacific
plate. Perhaps both processes could occur simultaneously.

At any rate, geologists know that during early Cenozoic time northern Honshu Island extended much farther into the Pacific than it does now. In 1978, for example, marine geologists reported that "remnants of a land mass that subsided during the current episode of subduction were found on the continental slope and on the upper part of the present Japan Trench's landward wall."[3]

Why would descent of the Pacific lithospheric plate beneath northern Japan suddenly accelerate? Owing to the pressures applied to Earth's plates due to pole shift and/or the upheavals in the mantle alluded to in the readings, starting around 1998, the slab descending beneath Japan suddenly might break through the 670 km barrier (Fig. 2) into the lower mantle in an "avalanche" of episodic mantle mixing of the type envisioned by some computer modelers.[4]

As the rate of subduction of Honshu and Hokkaido accelerates, so also will there be an increase in the frequency of earthquakes, tsunamis, and volcanic eruptions in and around these islands. Already in the pipeline, so to speak, are great destructive quakes for Tokyo and for the Tokai region some 70 miles southwest of downtown Tokyo. (See Fig. 17.)

The great (M 7.9) Kanto-district earthquake of September 1, 1923, killed 140,000 people in and around Tokyo. The upheaval, and the aftershocks and fires that followed, transformed the metropolitan area into a heap of ash and twisted metal. It was Japan's most destructive quake ever, until that time.

That was 73 years ago. The Tokyo region has suffered an earthquake of magnitude 8 or stronger every 69 years on average, according to a study of seismic activity since the year 818. These have been quakes originating directly beneath the metropolitan area or powerful tremors set off by earthquakes in neighboring districts. And legend has it that Tokyo will have a big quake five to ten years after an eruption of Mihara volcano, which lies about 65 miles south of downtown Tokyo. Mihara last erupted in 1986.[5] About 25 percent of Tokyo's soil is calculated to liquefy in the next large quake, causing many structures to fail. Firestorms following the shock are the biggest fear. Computer models of the hypothetical storms show that one-third of Tokyo will burn to the ground and 40,000 will die.[6]

In 1988, Kenshiro Tsumura, head of seismology at the Japan Meteorological Agency, said that he was not as worried about predictions of another Kanto district quake in the very near future as he was about the imminent possibility of a great quake in the neighboring Tokai region (Figs. 17 and 19). The last great Tokai quake, in 1854, was a magnitude 8.4 shock that killed upwards of 3,000 people and destroyed 25,000 homes. At the time, Tokai was a rather sparsely populated farming district, whereas today it is populated by about 7.5 million people living in over 170 towns and cities. It is also home to numerous oil refineries, industrial complexes, and petrochemical plants. Seismic forces have for several decades been building on faults beneath Suruga Bay, where the next Tokai-earthquake epicenter is expected to occur.[6]

Katsuhiko Ishibashi, one of Japan's top earthquake experts, believes the Tokai quake will trigger a series of geological events beneath Tokyo that would be powerful enough to destroy large parts of the city. Ishibashi's views were reported in the *Chicago Tribune* as follows:

> **His argument is based largely on historical evidence suggesting the 1854 Tokai earthquake touched off violent tremors in the Kanto District that caused numerous fatalities and great damage in**

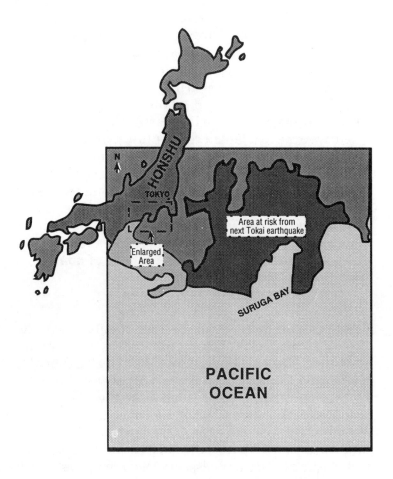

Figure 19
Tokai region. Japanese scientists have heavily instrumented the area around Suruga Bay to detect precursors of the next earthquake. (See Figure 17 for location of Tokai relative to plate boundaries.)

the years leading up to the catastrophic con-
vulsion of 1923. Because Tokyo has grown so
much in geographic size and population density,
he said, similar tremors today would be deadlier.

"When the great Tokai earthquake does occur, it
will plunge the Tokyo area into a period of intense
seismic activity marked by the repeated occurrence of magnitude 7-class earthquakes," he recently wrote in a Japanese scientific journal.

What does that mean in layman's terms?

Ishibashi, sitting in a 28th-floor restaurant over-
looking downtown Tokyo, swept his hand to take
in the sprawl of office towers, hotels, residences
rising above congested streets in what has often
been called the world's most poorly planned city.

"Collapsed buildings. Fires. Millions of panick-
ing people. Widespread death.

"Chaos," he said.[6]

Any sudden increase in the rate of subduction of
plates under Japan will bring terrible chaos to the population there. Consider the effects of the Kobe earthquake
of January 17, 1995. This was the most powerful tremor
to strike an urban area of Japan since the 1923 Tokyo
earthquake. The human toll of the Kobe quake was also
second to Tokyo's with over 5,000 dead and 20,000 in-
jured. In economic terms, the Kobe earthquake was the
most costly ever natural disaster. The tremor destroyed
or damaged more than 46,000 buildings, toppled the
main elevated highway between Kobe and Osaka, cut
gas, water, and electric service, derailed seven trains, and
left 300,000 of the city's 1.5 million people without shelter. Such chaos may be but prelude to the catastrophic
losses that would be expected due to subduction of the
greater portion of the Japanese archipelago itself.

6

Shaky Latin America and the Southern Hemisphere

Some of the worst natural disasters in modern times have occurred in Latin America. Earthquakes and volcanic eruptions have bypassed few countries in Central and South America. And more can be expected. As of this writing, May 1996, Mexico's El Popo (Popocatepetl), a three-mile high volcano southeast of Mexico City, had been forming a lava dome and ejecting glassy material and ash for 17 months. In December 1994, a minor eruption caused Mexican officials to evacuate 75,000 people from the volcano's eastern flank, the most probable site of disaster. Scientists are trying to determine whether new magma is rising within the mountain, signaling a more dangerous eruptive phase. Prehistoric ash falls have extended into what is now Mexico City.

Popocatepetl is but one of five active volcanoes in Mexico that may be expected to erupt after 1998, according to the Cayce readings, due to "upheavals in the Arctic and in the Antarctic." Other Central American countries that may expect similar volcanic action include Guatemala, which has nine volcanoes; Nicaragua, which has five; and Costa Rica and El Salvador, three each.

Major earthquakes have also struck this region in recent decades. These provide an idea of what might be expected in the near future:

• In the Michoacan, Mexico region 9,500 died in an 8.1 M event in September 1985. Mexico City suffered more than $3 billion in property damage, 30,000 inhabitants were injured, and a small tsunami was generated.

• In Guatemala, 22,778 died in a 7.9 M quake in February 1976.

• In the Managua, Nicaragua, region, 5,000 died from a 6.2 M quake in which a fault ruptured for over 200 kilometers in December 1972. Twenty years later Nicaragua was shaken as the result of subduction of the Cocos plate beneath the Caribbean plate, a quake that generated tsunamis that ran as high as 32 feet in places and caused extensive damage along the Pacific coast.

• In Costa Rica and Panama significant earthquakes could occur from ruptures in the Middle American Trench, offshore to the west of these countries, in the not-so-distant future. Recent studies of strain accumulation across the subduction interface at the Middle American Trench between the Cocos and Caribbean plates indicates that at least one segment of that interface is currently locked and is accumulating elastic strain (T. H. Dixon, *Geophysical Research Letters,* v. 20, Oct. 12, 1993). Large subduction zone earthquakes cluster in time rather than exhibit periodic behavior.

• In Mexico a 7.6 earthquake on October 9, 1995, rocked the state of Jalisco, a west-central region which in 1932 had Mexico's largest historical earthquake. In a paper entitled "Anticipating the Successor to Mexico's Largest Historical Earthquake" (*EOS*, v. 76, Oct. 17, 1995), seven geophysicists discuss that M-8.2 quake and 7.8 aftershock that caused widespread casulties and damage. As their paper was being prepared for publication, the 1995 quake hit that very region, in which Guadalajara is the largest city, again causing significant loss of life and property damage. Such quakes are related to ruptures in the Rivera subduction zone, offshore in the Pacific, which accommodates downward thrusting of the Rivera plate beneath the Jalisco region.

As for South America, reading 3976-15 predicts that it "shall be shaken from the uppermost portion to the end," that is, from Colombia and Venezuela in the north to the Chile-Argentina tip of the continent in the south. A potential for significant shaking and damage to property is quite expected, given the long history of pronounced seismicity in western South America (see Fig. 20). For a flavor of the kind of ongoing seismic shaking that has occurred in northern, western, and southern South America we need only look at the earthquake record for the period 1958-1998, when the readings said that the stage was being set for truly significant Earth changes ahead.

In 1959, a large quake took place in the northern Magellan Straits near the southern tip of South America. "This is unusual," Jim Lander of the U.S. Coast and Geodetic Survey told Science Service, "since there has never been an earthquake this far north in the Straits." Lander said that the quake was a deep one and quite strong. It was located at about 72 degrees west and 51 degrees

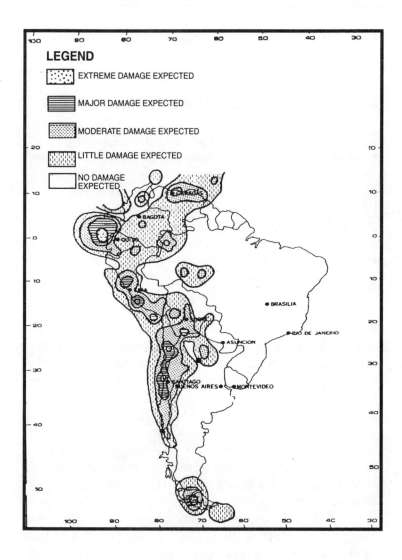

Figure 20
South American seismic damage map. Adapted, by permission, from a
map published by Wyle Laboratories, El Segundo, California.

south. This event coincides with our expectation from the portion of reading 3976-15 that says: "South America shall be shaken from the uppermost portion to the end, and in the Antarctic off of Tierra del Fuego *land*, and a strait with rushing waters."

The emergence of land will be signaled by seismic disturbances. Although Tierra del Fuego, at the tip of South America, is seismically active (see Fig. 20), the 1959 quake was interesting in that it was deep and it filled a gap in a chain of quakes that extended from 47 to 53 degrees south latitude.

The next big quakes in South America occurred in 1960, over a nine-day period (May 21-30) in southern Chile. The largest quake was no less than an 8.5 M event that generated a large tsunami. Some 5,700 people died. This was followed in March 1965 by a quake near Santiago, Chile, which killed 400 and produced $200 million in damage. In July, 1967, an earthquake that shook Venezuela was felt from the Andes to the Caribbean. This quake in far northern South America killed 277 people and caused $15 million in damages in Caracas.

None of these disasters can compare with what happened a few years later in Peru. In October 1969 quakes leveled two villages. But the destruction there was but a precursor of horrors to come. Eight months later, on May 31, 1970, halfway through a quiet Sunday afternoon, the earth began to tremble in a high mountain valley of north-central Peru. Before it stopped, a mighty earthquake registering 7.7 on the Richter scale had shaken the Andes, shattering town after town in a long narrow valley known as the Callejon de Huaylas. Some 75,000 people perished in the quake and in the ice and mud avalanches that it spawned, making it by far the worst natural disaster in the history of the Western Hemisphere. And despite Peru's history of repeated temblors, the 1970 catastrophe took more than double the toll in lives of all of the country's previous five centuries of earthquakes. In "No Bells to Toll," which brings that extraordinary event in the Peruvian Andes to life, anthropologist Barbara Bode quotes a Peruvian who experienced that calamity:

> **This is not the terror felt by someone running from a furious bull, a threatening knife ... This is a cosmic terror, an instant danger, the universe caving in and crumbling away. And, meanwhile, the earth lets out a muffled sound of thunder, in a voice no one knew it had. The dust raised by the houses as they came crashing down settles little by little. And we are left, with all the dead, not knowing how we happen to be still alive.**
> **—Pablo Neruda, *Memoirs***

The sudden devastation of this earthquake in Peru reminds me of the following Cayce reading:

> **In the one [incarnation] before this we find in that day when there was the destruction to the elements of earth [destruction of Atlantis?], then in now the *Peruvian* countries, when the lands disappeared in the low places and the entity was left alone in the higher mound to which the entity had then gone for its study. In the name then of Oumu. In the personality exhibited in the present, the ever fear in the deep recesses of self of destruction coming to many suddenly.**
> **182-2**
> **April 9, 1925**

Western Argentina was jolted by a powerful quake in

November 1977. About 70 died, and at least 10,000 were left homeless. The tremors were felt over a wide area of Chile, Peru, and Brazil as well. On December 12, 1979, an M 7.7 quake hit near the coast of Ecuador killing 800 and injuring 20,000. A 10-foot tsunami devastated the entire southwest coast.

On March 3, 1985, a powerful M 7.8 quake hit the Valparaiso, Chile, region, killing 177 and injuring 2, 575. Then, on July 30, 1995, a strong (M 8.1) subduction quake hit the Antofagasta (North Chile) area, one of the strongest there in a century. Ecuador was hit again in March 1987, this time in the northeast, by quakes that killed over 4,000. Northern Peru was struck by an M 6.3 earthquake on May 30, 1990, where 115 perished.

Seismic activity seemed to intensify in 1995. In January a strong (M 6.5) quake in eastern Colombia's sparsely populated plains rocked Bogota and was felt as far away as Caracas. In February the most deadly quake in Colombia in 12 years hit about 130 miles southwest of Bogota. In Pereira, it toppled tall buildings and opened huge cracks in the earth. In July, the strongest earthquake in decades struck Chile's northern copper-mining region.

Ecuador was shaken once again in March 1996, when approximately 62 people perished and hundreds were made homeless by a devastating quake. As bad as the shaking of South America has been down through the ages, I think that reading 3976-15 may be describing an *extraordinary* type—or amount—of earthquake shaking during the years of Earth changes immediately ahead.

Such a phase of earthquake activity may have begun with the June 9, 1994, magnitude 8.2 quake that occurred beneath Bolivia, at the great depth of nearly 400 miles. Minutes after the quake, strong ground motion was felt over most of South America. So powerful was this deep quake that it was felt throughout much of the Western

Hemisphere. It even excited modes of vibration of the entire Earth never before seen. Such deep quakes take place in the easterly descending slab of Pacific Ocean crust beneath western South America. But because of the more forceful western movement of the overlying continent of South America, the descending slab is itself moving west. It is being pulled "through the ductile rock of the mantle like a garden hoe through soil . . . [and] . . . the mantle diverges like fine soil around [the] hoe, flowing north and south along the west face of the slab."[1] The flow is disrupted, however, where it changes its descent angle and a cluster of former Bolivian deep quakes has been determined by seismologists to lie at the bottom of one of the slab wrinkles. Any increase in speed of movement of the slab or of the westward-moving continent, will cause an increase in continued shaking of South America from "the uppermost portion to the end." Thus, we can expect more great earthquakes in western South America in the near future.

As to the timing of the onset of the first major episode of increased shaking in the Southern Hemisphere, consider again reading 270-35:

> **If there are the greater activities in the Vesuvius, or Pelée, then the southern coast of California— and the areas between Salt Lake and the southern portions of Nevada—may expect, within the three months following same, an inundation by the earthquakes. But these, as we find, are to be more in the Southern than in the Northern Hemisphere.**
>
> **270-35**
> **January 21, 1936**

South America, the largest of the continents of the Southern Hemisphere, also has more active volcanoes than North America. During the 1958-1998 period that Cayce referred to, records show the following eruptions:

- In Ecuador: Cotopaxi (1975), Guagua Pinchincha (1981, 1982, and 1989), and Sangay (1976 and 1988).
- In the Galapagos: Alcedo (1970 and 1979).
- In Colombia: Purace (1977), Nevado del Ruiz (1985, 1986, 1987, 1988, and 1989), and Galeras (1990). Galeras has been restless since 1988 and looms over the large city of Pasto. In 1993, an unexpected, small explosion killed six volcanologists on a field trip into the cone.

All of these are Torrid Zone volcanoes. Most active of the non-Torrid Zone volcanoes is Chile's Villarrica, showing continuous lava-pond activity since 1985.

As for other danger areas in the Southern Hemisphere, earthquakes occur throughout New Zealand because of the country's location astride one of the world's major plate boundaries. Beneath the North Island the Pacific plate is being subducted under the Australian plate. Southwest of the South Island, the Pacific plate (including the South Island) is overriding the Australian plate. Between these two opposing subduction systems, the plate margins are in oblique collision, and the New Zealand landmass is being twisted and torn by complex horizontal faulting and vertical movements. [Incidentally, this currently slow "breaking up" of New Zealand is taking place in an area of the South Pacific that's *almost* opposite to the Mediterranean and the Etna area, as discussed at the beginning of Chapter Three. It's slightly closer than the Kermadec-Tonga trench (KTT) that we said earlier was the most probable location of the South Sea breakup, mentioned in reading 311-8. The KTT is the most probable location because it's the locus of the Earth's most active zone of *mantle* seismicity.] Historical earthquakes in New Zealand of greatest importance are the major disturbances of 1848, 1855, 1929, and 1931, all of which showed evidence of faulting. The 1848 quake, for example, was perceptible over about half of the country, was violent on both sides of Cook Strait, and resulted in fissures and fresh scarps in the Awatere Valley. Wellington, the capital, was seriously damaged in the quake of 1855. The country to the west was generally uplifted from three to ten feet. The Hawke Bay earthquake of February 3, 1931, was the first quake disaster in the country, killing 255. Most of the material losses were due to shaking and fire at Napier and Hastings. The wharf at Napier was uplifted nearly seven feet. Numerous other examples of earthquakes and associated faulting (up or down) could be cited.

Relative movements between the plates beneath New Zealand have produced a rugged topography and active volcanoes. North Island's Ruapehu volcano exploded spectacularly on September 23, 1995. Ruapehu is the major ski resort on the North Island, attracting 10,000 skiers on some days. But the eruptions that continued into October 1995 produced lahars, emptied a large lake, and dispersed a large volume of ash and scoria bombs downwind.

The tectonic setting of Christchurch, "the garden city" on the South Island, indicates that future large earthquakes are very likely to occur that will have a serious impact on the city, even if local plate motions are not accelerated in a Cayce-readings-type Earth-change scenario. But if the plates do speed up, all of New Zealand will be greatly impacted by quakes, eruptions, and vertical movements of the crust.

We might expect severe shaking in New Zealand, South America, and Antarctica, as a pole shift gets underway and after Vesuvius and Pelée begin to show increasing activity. The Indonesian region will also probably be severely affected both by seismic shaking and by the eruption of its many Torrid Zone volcanoes. Australia and southern Africa on the other hand are relatively far removed from plate boundaries and have lesser inherent seismotectonic potential.

7

Lost Continents

Is the "lost continent" of Atlantis a poetic myth or a part of the turbulent prehistory of our planet, as Cayce and others suggest? Philosophers and writers worldwide, from Plato in ancient Greece to Zhirov in modern Russia, have spun intriguing tales of Atlantis without certain evidence of its existence. Cayce claimed that at least one other continent, Lemuria, in the Pacific, also vanished. This mystery of "lost civilizations" captivates our imagination and continues to inspire explorers, adventurers, even scientists like myself to search for clues, to fashion a plausible theory that would suggest locations and explain their disappearance.

But first, how could continents appear and disappear

in the ocean basins? What forces could be at work beneath the crust that would be able to elevate wide areas of the sea floor above the ocean surface? Having no hard evidence for continental-scale emergence of ocean floors, scientists have no theories that explain how this could happen. Instead, geoscientists are busy explaining the morphology of ocean floor features, island arcs, continental shelves and trenches, and the motions of continents through time by means of the theory of sea-floor spreading. Before even citing any readings that refer to lost continents rising out of the ocean, we must engage briefly with this theory that has become the model for the "new global tectonics."

Sea-floor spreading occurs when hot rock (basalt magma) upwells from beneath the ocean floor from the upper mantle and forms a new ocean-floor layer at an oceanic ridge (see Fig. 9, for example). As the magma cools, magnetic minerals in the new material take on the same polarity as that of the magnetic field of the Earth at the time of crystallization. When the Earth's magnetic field periodically reverses its polarity, any new basaltic lava forming at the ridge acquires the reversed polarity magnetization. Sea-floor spreading is believed to take place symmetrically, older lavas with their recognized magnetic polarities being carried away equal distances on either side of the ridge. This produces a symmetrical pattern of magnetic stripes about the ridge axis. Thus, the crust of the ocean floor is said to contain a magnetic record of its own formation. Scientists have numbered the magnetic lineations from 1 at the ridge to about 190 near the continental margins. These magnetic patterns suggest that the world's ocean floors were all created since the Jurassic period, about 180 million years ago, and that the continents must have moved apart in response to continuing formation of new ocean crust. New crust is formed by magma conveyed to the ridge crests by convection cells in the mantle. Note that there is nothing in the theory of sea-floor spreading—as it is currently practiced—that supports the idea of continental-sized pieces of the oceanic crust having once stood above the surface of the ocean; continents like Atlantis and Lemuria, that is.

The theory of sea-floor spreading began with the discovery of linear magnetic patterns in the rock of the ocean floor, as measured from ships and planes over the last 40 years or so. The patterns seemed to show a general symmetry of positive and negative magnetic anomalies that paralleled a central axis, such as the mid-ocean ridge system that circles the globe. Geophysicists in 1963 concluded that the magnetic anomalies were once connected and had spread apart. The idea was then advanced that new ocean crust is continuously generated along an ocean midline, or axis (mid-Atlantic ridge, say) and pushes older crust to the sides. The linear positive and negative anomalies that parallel the ocean midlines are thought to have been caused by newly formed basaltic crust cooling and becoming magnetized in a field that reverses itself at various times, imprinting in this manner successive patterns on the cooling iron-rich basalts that have been exuded from the roughly linear mid-ocean ridge. Offsets in these patterns were found to correlate with faults running perpendicular to the mid-ocean ridges. Investigation of the faults showed that they were a special type that reflected sea-floor spreading in directions away from the ridges. This was the development of the transform fault[1] hypothesis, and it led to a bandwagon of geologic interpretation and opinion that "convection cells" beneath the oceanic crust bring hot rock up into the mid-ocean ridges and explain all the observed phenomena just mentioned. Ultimately, the continental drift theory from Alfred Wegener's time was revived and the great "plate race" was on.

One of the early critics of the new global tectonics, H.J. McCunn, wrote:

> **From the real data—the magnetic patterns—it has been rather a short trip through three major, separate hypotheses (magnetic reversals, sea-floor spreading, and convection cells) to the ultimate theory of "drift." In science, it is dangerous to stack hypothesis upon hypothesis . . . All of the phenomena named above have alternate explanations that are equally good and, in some cases, superior, because they call for no large amounts of unexplained energy; i.e., they can be explained in the light of what is observed. Highly zealous proponents of the new global tectonics claim one must believe in the concepts because they explain so much . . . However, [A.A.] Myerhoff and [H.A.] Myerhoff . . . concluded that the new global tectonics explained very little and reviewed the major inconsistencies in the concepts. [2]**

McCunn went on to propose a model of plate tectonics based on vertical uplift, cooling, and collapse, driven by Earth's internal heat, gravity, and isostatic compensation. (This last involves adjustment of Earth's lithosphere to maintain equilibrium among units of varying mass and density. Excess mass above is balanced by a deficit of density below, and vice versa.) McCunn's paper,[2] published in 1973 and largely ignored since then, is resurrected here to show that alternative explanations may hold for tectonism in some parts of the ocean basins. The sea-floor spreading model currently holds sway among geodynamicists, and it *has* led to a unifying theory of plate tectonics that, as presently practiced by geoscientists, seems best to explain how the Earth works. But vertical uplift, cooling, and collapse of portions of the oceanic crust may also have a place in understanding the evolution of the ocean basins. We will assume here that vertical tectonics were behind the emergence, cooling, and then sinking of the continents of Atlantis and Lemuria, in the Atlantic and Pacific Oceans, respectively, during the past 10 million years. There is but one special feature in this assumption. It is that the breakup and sinking of Atlantis was hastened by human error, as the Cayce readings claim.

THE DESTRUCTION OF ATLANTIS

The Cayce readings explained how Atlantis was destroyed over several thousands of years:

> ... [men on Atlantis] brought in the destructive forces as used for the peoples that were to be the rule, that combined with those natural resources of the gases, of the electrical forces, made in nature and natural form the first of the eruptions that awoke from the depth of the slow cooling Earth, and that portion [of Atlantis] now near what would be termed the Sargasso Sea first went into the depths.
>
> 364-4
> February 16, 1932

Q Describe in more detail the causes and effects of the destruction of the part of Atlantis now [known as] the Sargasso Sea.

A ... individuals added to that used in the form of what is at present known as the raising of the powers from the sun itself, to the ray that makes for disintegration of the atom, in the gaseous forces formed, and brought about the destruction in that portion of the land now presented, or represented, or called, Sargasso Sea.

Q What was the date of the first destruction, estimating in our present day system of counting time in years B.C.?

A Seven thousand five hundred (7,500) years before the final destruction, which came as has been given.

364-11
April 29, 1932

The use of these influences by the Sons of Belial brought, then, the first of the upheavals; or the turning of the etheric rays' influence *from* the Sun—as used by the Sons of the Law of One—into the facet for the activities of same—produced what we would call a volcanic upheaval; and the separating of the land into *several* islands—five in number.

Poseidia, the place or settlement of that particular sojourning of the entity—Deui—at the time, then became *one* of these islands.

877-26
May 23, 1938

The principal islands at the time of the final destruction of Atlantis were called Poseidia, Aryan, and Og. These islands resided in an archipelago that would fit within the boundaries of the present Sargasso Sea, or *near* those boundaries, and were submerged some 11,900 years ago in the final destruction of Atlantis.

Q Give in detail what the sealed room contains.
A A record of Atlantis from the beginnings of
those periods when the Spirit took form or began
the encasements in that land, and the developments of the peoples throughout their sojourn,
with the record of the first destruction and the
changes that took place in the land, with the record
of the *sojournings* of the peoples to the varied activities in other lands, and a record of the meetings
of all the nations or lands for the activities in the
destructions that became necessary with the final
destruction of Atlantis and the buildings of the
pyramid of initiation, with who, what, where,
would come the opening of the records that are as
copies from the sunken Atlantis; for with the
change it *must rise (the temple) again* [emphasis
added].

378-16
October 29, 1933

Before that we find the entity was in the Atlantean
land, when there were the constructive forces as to
the activities of the children of the Law of One—in
all of those influences during the periods when the
land was being broken up. We find the entity was
as the leading influence for the considering of
ways and means in which there would be the preserving of records, as well as ways, means and
manners in which either the few or the numbers
might be preserved from the destruction of the
lands. It would be well if this entity were to seek
either of the three phases of the ways and means
in which those records of the activities of individuals were preserved—the one in the Atlantean land,

that sank, which will rise *and is rising again* [emphasis added]; another in the place of the records that leadeth from the Sphinx to the hall of records, in the Egyptian land; and another in the Aryan or Yucatan land, where the temple there is overshadowing same.

2012-1
September 25, 1939

To summarize, the continent of Atlantis was reduced to five islands in a man-made volcanic cataclysm around 19,400 years ago. It was further diminished from five islands to three islands, one of which was Poseidia. The final destruction occurred over an 800-year period between about 12,700 to 11,900 (288-1) years ago and was a direct consequence of the wicked nature of a preponderance of the remaining inhabitants. The temple on Atlantis that contains the records of that continent's existence is said to be "rising again." But where is this temple located? Beneath the Sargasso Sea somewhere? Or in the vicinity of the Bimini Islands, in the northwestern Bahamas? We'll search for the answer a little later on.

Is Cayce's story of Atlantis credible? We can only come to a tentative conclusion because, at least until 1998 has come and gone, it depends upon the credence one gives to Cayce's way of knowing. His was not the ordinary logical way of explaining life by making observations, collecting facts, and inferring laws from the process. Nor was it just the intuitive way, in which one sees on rare occasions, via a special form of inspiration, that the whole phenomenal world is the product of laws. No, Cayce could not only read the subconscious minds of others, he could access the scroll of records found at the junction of time and space, the so-called akashic records, and he could obtain spiritual and other commentaries on the information obtained there as well. In the trance state he was able to access the universal consciousness, the soul minds of higher masters, and other sources.

As geoscientists probe ever more deeply into the "record in the rocks," their findings will oftentimes coincide with the Atlantis story of the Cayce readings. For example, a recent study[3] of the concentrations of calcium (Ca^{2+}) and sulfate (SO_4^{2-}) ions in annual ice layers cored from the Greenland ice sheet provides a 110,000-year record of explosive, high-sulfur-producing volcanism that shows a strong correlation with Atlantean "disturbances" and eruptive events mentioned in the readings.

As an illustration, some 52,682 ice-layer years ago, the sulfate-ion value in the GISP2 ice-core record is one of the largest of the 838 volcanic signals in the entire ice-layer sample series. And this signal, representing enormous volcanism somewhere in the Northern Hemisphere, is followed by around 1,000 years of calcium-ion peaks that are a primary indicator of continental dust. The date of the eruption signal coincides almost exactly with the 52,718 B.P. date given in reading 5249-1 for the gathering of concerned individuals to develop a plan to rid the Earth of enormous animals, as described in Chapter Two. In addition, a volcanic ash layer found in a different (GRIP) ice core from Greenland and believed to have originated from an area of explosive volcanism in southern Iceland, is found to be 52,700 ice-core years old.[4] This information could well be describing the beginning of my hypothesized movement of the North Pole from some unknown position elsewhere on Earth to northern Greenland, where it stayed until 19,400 B.P. (see Chapter Two). Then, when the continental ice sheets had expanded to their limits in the Northern Hemisphere, and men on Atlantis tuned the firestone too high and "produced what we would call a volcanic upheaval" (877-26) that split Atlantis into five islands, the pole shifted again to its present position.

At 19,400 B.P., however, the chronology of Atlantean destruction seems to lack much correlation with the volcanism signals from the ice cores. The closest sulfate-ion signal to 19,400 B.P. in the GISP2 core is one at 19,058 B.P. This lack of significant sulfate-ion signals around 19,400 may be due to the unique nature of the man-made eruption. It may not have been a high sulfur-producing eruption, the kind necessary to produce a sulfate-ion signal. Or the volcanic activity could have been largely submarine in nature, thereby releasing little in the way of sulfate ions into the atmosphere. It's also possible that the sampling frequency of the ice layers in GISP2 was not fine enough to detect the 19,400 B. P. volcanic catastrophe inferred from the readings.

The GISP2 ice-core record between 31,000 and 28,000 B.P. is intriguing. It provides evidence[3] of a period of significant volcanism that occurred somewhere in the Northern Hemisphere (Atlantis?) at the same time that reading 470-22 said that a second period of "disturbances" were underway on Atlantis. The time when "the entity" was experiencing these disturbances was given as 30,000 B.P., and the disturbances were so strong that "small channels [were produced] through many of the [Atlantean] lands."

Perhaps most significant of all of the ice-core records of volcanism is the finding[3] of a discrete and pronounced record of well-above-baseline peaks in volcanic signals for the period between 12,657 and 11,285 B.P. This broad and strong peak in sulfate concentrations describes a significant period of explosive eruptions that matches almost exactly the interval given in the readings for the final 800-year-long destruction of Atlantis. That period began in 12,700 B.P. (364-4) and ended in 11,900 B.P.

(288-1). And the ash layer found in the Greenland GRIP ice core, at 11,980 (±80) ice-core years ago, corresponds to a great eruption that deposited Ash Zone 1, a significant marker horizon in North Atlantic sediments.[4] A fitting epitaph for the end of Atlantis!

Returning to the question of the location of the "lost continent," which the Cayce readings place near the Sargasso Sea, the *Encyclopaedia Britannica* tells us that the Sargasso Sea is "a tract of the North Atlantic Ocean covered with floating seaweed (*Sargassum ...*) ... bounded approximately by 25° and 31°N and by 40° and 70°W, but its extent and density are influenced by winds and ocean currents." The northwestern corner of this tract lies in the ocean southeast of Cape Hatteras and the southwestern corner lies to the northeast of the southern Bahamas. The eastern side of the tract lies due west of the Canary Islands and far out into the Atlantic near the Mid-Atlantic Ridge.

Instead of using the boundaries just quoted, it seems best to use an average of several latitude and longitude boundaries for the Sargasso Sea, as given in five different sources. The average boundaries used here will define the area of the North Atlantic Ocean bounded by 23° and 35°N and 35° and 72°W. Maps of this area, found in Volume M of the Decade of North American Geology (DNAG) study,[5] show highly varied features of the ocean floor.

Proceeding from east to west, this area includes the eastern flanks of the Mid-Atlantic Ridge (MAR), the MAR itself from the Kane Fracture Zone on the south to the Oceanographer Fracture Zone on the north, the Muir Sea Mount, the Kane Fracture Valley, a small part of the Hatterras Abyssal Plain, the Bermuda Rise and Bermuda proper, the Nares Deep, the Nares Abyssal Plain, and the Vema Gap.

In reading 364-4: we read: " . . . and that portion [of

Atlantis] now near what would be termed the Sargasso Sea first went into the depths." The word *near* is important. But does it mean *near* to the east, west, north, or south sides of the boundaries of the Sargasso Sea?

The best scientifically oriented study of the Atlantis hypothesis I know of was published in Moscow in 1970 by the Russian doctor of chemical sciences, N.F. Zhirov. A 437-page English translation, entitled simply *Atlantis,* shows that Zhirov made an exhaustive analysis of then known oceanographic, geologic, and biologic information and concluded:

. . . there are grounds for assuming that in the sinking of Atlantis there were two stages, the first between the 13th and 10th millennia B. C., and the second, the most considerable, between the 9th and 8th millennia B.C.

Zhirov's map (Fig. 21) locates Atlantis between 24° and 45°N, and 03° and 47°W. It shows boundaries for continental fragments of Atlantis (1) at the time of the surmised maximum extent of Atlantis (along the crest of the MAR, along the continental shelves of Spain and Morocco, and around the shallower portions offshore of the Canary Islands, Madeira, Tin Island, and the Erytheia archipelago); (2) before the main subsidence (includes the present Atlantis and Plato sea mounts, one small and one very large region of the MAR); and (3) after the main subsidence (which includes the Azores and associated offshore areas of the MAR).

If Zhirov is correct, the area *near* the Sargasso Sea that sank first would be both to the east and to the west of 35° W longitude, our eastern boundary for the Sargasso Sea. This seems close enough to being *near* the Sargasso Sea to satisfy the requirement of the Cayce reading. And in-

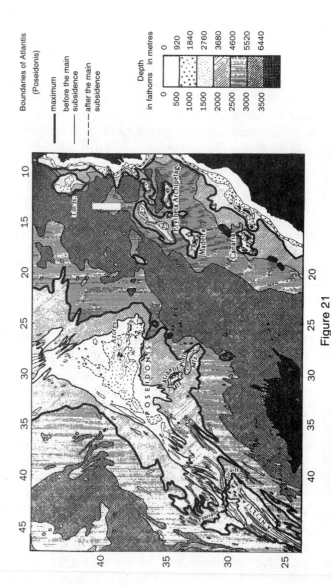

Figure 21

N. F. Zhirov's figure entitled "Surmised Location of Atlantis between 25°N and 45°N." Zhirov coined the names Poseidonis, Antillia, Poseidon's Straits, Atlantis Mts., and Plato Mts. Today's marine geologists have named Zhirov's Atlantis Mts. the Atlantis Seamount, his Plato Mts. the Cruiser Tablemount, and the elliptically shaped feature almost due south of Zhirov's Plato Mts., the Great Meteor Tablemount.

terestingly, Zhirov's final destruction of Atlantis in "the 9th and 8th millennia B. C.," is rather close to the readings' period of final destruction between 10,700 and 9,900 B.C. Zhirov's date of the 13th millennium B.C. for the destruction of the main part of Atlantis is far off from the readings' 17,400 years B.C., however.

All things considered, Zhirov's outlines of Atlantean land areas "before the main subsidence" (Fig. 21) are probably the best that science can provide by way of "corroborating" Cayce's Atlantis readings.

But if Zhirov is right, then why did the readings continually locate sunken Atlantis in an area of the ocean near the Sargasso Sea? Why did the readings not just say "southwest of the present Azores" or some such? And what could be the location of the man-induced volcanic cataclysm that the readings indicated had split the Atlantean land into five islands only 19,400 years ago? Could it have occurred at what scientists have identified as the Azores hot spot, a mantle plume located at about 38.5° north and 28.0° west? Or has the thermal activity of the man-induced volcanic upheaval disappeared and the associated sea-floor morphological evidence become indistinguishable from the surrounding fractured, lava-flow-covered terranes of the ocean bottom beneath the Sargasso Sea? Only if marine geoscientists look specifically for evidence of the cataclysm described in the readings will we ever begin to know the "where and when" of this watershed event in the history of Atlantis.

But, you say, scientists have no time to waste on such problematic research. They've got explanations for everything they see on the ocean floor right now, as explained by the theory of sea-floor spreading. That's not completely correct, as evidenced by a review published in *Nature* (11 April 1996, p. 480) on "anomalous" findings in the Romanche fracture zone near the Mid-Atlantic Ridge in the equatorial Atlantic. Here, sediments 140 million years old lie in part of the Atlantic Ocean where the crust is only a fraction of that age. That's not supposed to happen, according to the theory of sea-floor spreading, because such sediments are supposed to have been conveyed to the edges of the ocean basin by now, as new oceanic lithosphere is continuously formed at the mid-ocean ridge and pushed toward the continents on either side. And so how are geoscientists explaining this one? They are questioning their assumptions as to the stability of the transverse ridge where the sediments were found, and they are proposing instead that the transform faults that would normally convey-away the ridge sediments somehow "jump around," entrapping the old sediments near the ridge crest. The review article states flatly that "this discovery cannot be explained by conventional plate tectonics" and goes on to say that "the existence of transverse ridges has long puzzled Earth scientists. Their morphology is complex and resists generalization. Rocks retrieved from these ridges vary from upper mantle to shallow-water carbonates, and this variability is taken as an indication of the intense vertical tectonism that can occur."

My point here is that there is much that we don't yet know about tectonism on the ocean floors, and current interpretations may be just as capable of "jumping around" as transform faults might be. There's an old geologist's saying, "If I hadn't known what I was looking for, I wouldn't have found it." It is hoped that some marine geoscientists will look through the existing data on the North Atlantic Sea floor in the vicinity of the Sargasso Sea to see if there is any evidence for a land mass that partially foundered around 19,400 years ago after breaking up into five islands and for a volcanic upheaval that triggered this cataclysm.

But what of the *western* portions of the once great continent of Atlantis described in the readings? Note that the boundaries of the Sargasso Sea used above do not include the Bahamas and Bimini Islands. To the west of our western boundary of the Sargasso Sea lies the Blake Outer Ridge; farther to the west lies the Blake Plateau and the continental shelf. (See Fig. 22.) The famous marine geologist, F. P. Shepard, argued "that something in the nature of faulting accounts for the bulk of the continental slopes,"[6] and he noted that "south of Cape Hatteras there is a radical change in the character of the continental slope (see Fig. 22). One or more broad steps are found leading down to the deep ocean floor. Also, the base of the continental slope is no longer bordered by a continental rise. Instead, the steep slope beyond the terraces ends in a trough. From the latitude of South Carolina to the Bahamas there is one principal terrace, called the Blake Plateau . . . It has long been known that this plateau had little if any sediment cover. Outside the plateau there is an abrupt drop with a slope described as 50 percent in some profiles.[6] This description of the ocean floor lying to the west of Cayce's sunken Atlantis sounds like a step-faulted continental shelf, bounding the western margin of a large down-dropped area that is now the deep Atlantic Ocean floor. And the reading below sheds light upon what the Cayce source would have to say about the relationship of the Blake Plateau to Atlantis.

> **Q Describe the Earth's surface at the period of the appearance of the five projections [of the races, around 10 million years ago].**
>
> **A This has been given. In the first, or that known as the beginning, or in the Caucasian and Carpathian, or the Garden of Eden, in that land which lies**

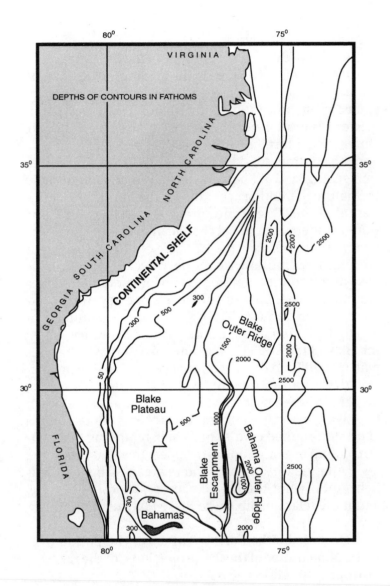

Figure 22
Physiographic features of the Blake Plateau. Note change in continental
slope at Cape Hatteras and the broad Blake Plateau to the south.

now much in the desert, yet much in mountain
and much in the rolling lands there. The extreme
northern portions were then the southern portions, or the polar regions were then turned to
where they occupied more of the tropical and
semi-tropical regions; hence it would be hard to
discern or disseminate the change. The Nile entered into the Atlantic Ocean . . . What is now the
central portion of this country, or the Mississippi
basin, was then all in the ocean; only the plateau
was existent, or the regions that are now portions
of Nevada, Utah and Arizona formed the greater
part of what we know as the United States. That
along the Atlantic board formed the outer portion
then, or the lowlands of Atlantis.

364-13
November 17, 1932

Extensive geophysical studies of the Blake Plateau by
W.P. Dillion and P. Popenoe[7] lend little support to the
above reading's assertion that "the lowlands of Atlantis"
were "along the Atlantic board." Only if the Blake Plateau
had been very briefly elevated, geologically speaking,
would it have been possible for the plateau to have taken
form in the way that geoscientists have worked out for
its developmental history.

Regardless of whether the Blake Plateau represents a
submerged portion of the lowlands of Atlantis, we notice that at its southern end it merges into the Bahamas.
And how do the Bahamas fit into the Atlantean continent of the Edgar Cayce readings?

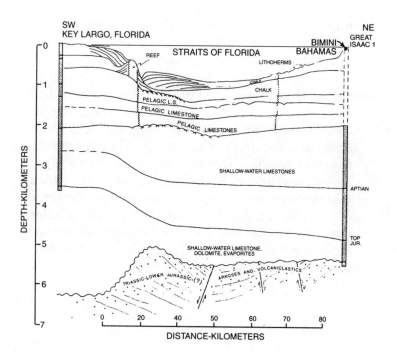

Figure 23
Simplified structural cross section of the northern Florida straits. Dashed part of Great Isaac Island oil-exploration well indicates that published accounts are only available for this part of well. Vertical exaggeration=20x. Adapted from Figure 10 of R. E. Sheridan, and others, 1981; reprinted by permission.[8]

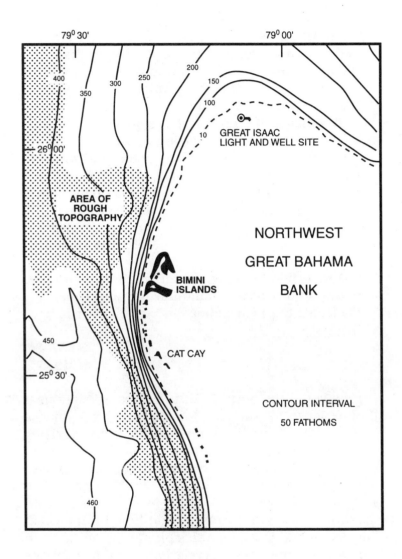

Figure 24
The Bimini Islands, showing bathymetry and area of rough bottom topography. Great Isaac Well 1 appears also, in vertical section, in Figure 23.

The position as the continent Atlantis occupied, is that as between the Gulf of Mexico on the one hand and the Mediterranean upon the other . . . There are some protruding portions within this that must have at one time or another been a portion of this great continent. The British West Indies or the Bahamas, and a portion of same that may be seen in the present—if the geological survey would be made in some of these—especially, or notably, in Bimini and in the Gulf Stream through this vicinity, these may be even yet determined.

364-3
February 16, 1932

And, as may be known, when the changes begin, these portions [of, as a minimum, the Great Bahama Bank near to and including Bimini] will rise among the first.

587-4
July 1, 1935

How could the portions of the Bahamas area near Bimini rise to the surface? Assume that the worldwide tectonic plates begin to move in response to rising mantle plumes and/or to horizontal mantle motions related to the forthcoming pole shift. Geophysicists have identified high-angle faults (see Fig. 23 for a partial structural cross section) in deeply buried rocks beneath the limestones and dolomites that hold up the Great Bahama Bank (Fig. 24) on which Bimini sits and beneath the Gulf Stream to the west (Fig. 23). The upper surface of the rocks is about three miles deep and the surface extends from roughly the midpoint of Great Bahama Bank westward across the Florida Straits to Florida. Thus, the entire area shown in Fig. 24 is underlain by these volcaniclastic rocks. According to sea-floor spreading theory, the high-angle faults in these rocks define blocks of the ancient rift crust that was emplaced when North America split away from Africa in the Middle Jurassic, between 135 and 190 million years ago. During any accelerated *westward* movement of the Atlantic lithospheric plate in the vicinity of the Bahamas, these faults could be reactivated and some of the blocks upthrust. But of possibly higher probability in terms of the readings, deeper mantle movements could result in a more direct uplift of the entire Bahamas area, including any underlying oceanic basement rocks.

K.D. Klitgord and H. Schouten, who analyzed plate motions of the central Atlantic, say that "long periods of constant spreading in the central Atlantic were interrupted by short periods of plate motion adjustment."[5] Some of these short periods were possibly times of emergence of portions of the ocean floor. An oncoming period of plate motion adjustment would seem to coincide with the following reading, cited earlier.

These changes in the Earth will come to pass, for the time and times and half times are at an end, and there begin those periods for the readjustments.

294-185
June 30, 1936

And Klitgord and Schouten conclude that "throughout this history it is the tectonic activity at either end of the Central Atlantic spreading center, the Iberian-Grand Banks segment, or the Gulf of Mexico-Caribbean segment, that has the most important geologic ramifications." This may relate to the reading that says:

There will be new lands seen off the Caribbean Sea, and *dry* land will appear.

3976-15
January 19, 1934

INVESTIGATING BIMINI

Reading 996-12 states that "a temple of the Poseidians was in a portion of this land" [around Bimini] and says that careful investigations in the Bimini area (see Fig. 25) will reveal this temple and other remains of the Atlantean continent.

Edgar Cayce's interest in the Bimini Islands was inspired by a friend for whom he had given readings after the man lost his eyesight in a car accident. The man was cured by following a healing program prescribed in his readings. After that dramatic success with Cayce, he sought advice from the clairvoyant for his business affairs, including many readings on his coal-mining operations in Kentucky and, later, several readings (the 5714 series) on oil wells in Florida. Knowing of Cayce's difficult financial condition and of his desire to build a hospital in Virginia Beach, this man appealed to Cayce to join a small group of businessmen interested in finding oil or buried treasure on Bimini. Cayce was to get a large fee for whatever his readings might successfully locate.

One member of this group was a multimillionaire who owned considerable property on North Bimini, including the Bimini Bay Rod and Gun Club and its marvelously appointed Hotel Bimini. When the club and hotel were destroyed in September 1926 by a hurricane that also devastated Miami, the owner suffered heavy financial losses.

Cayce gave his first reading on Bimini at his office in

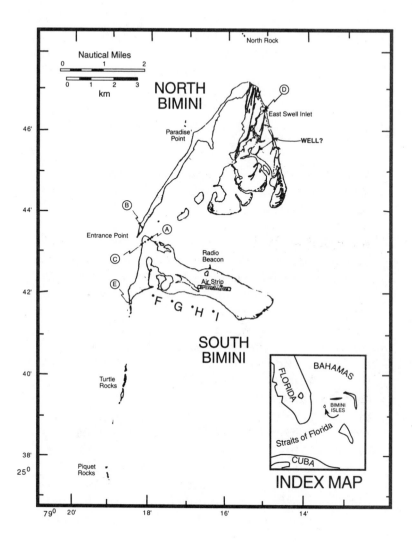

Figure 25
Map of Bimini Islands showing drilling and sampling sites, and features
mentioned in text.

Virginia Beach, one month before the hurricane hit. He was asked for information on the presence of oil and buried treasure there. Here are the main points of the reading that relate to Atlantis:

> **Dr. House: You will give a reading on Bimini Island, located in the Atlantic Ocean, about 45 miles almost due east of Miami, Dade County, Florida. You will go over this island, and tell us whether there is oil on this island in sufficient quantities to insure profit. If so, give us the log of the formations that would be gone through in drilling to the oil sand, depth of the well necessary to reach the oil production, also tell us if there are any treasures buried on this island. If so, when were they buried, and where can they be located, and to whom do they belong at this time?**
>
> **Edgar Cayce: Yes, we have the land known as Bimini, in the Atlantic Ocean. In the formation of the land, we find this of the nature that would make the oil production very low, for this is of the coral structure in the greater part, but this is the highest portion left above the waves of once a great continent, upon which the civilization as now exists in the world's history found much of that as would be used as means for attaining that civilization.**
>
> **Then, as to treasures, there are many that are hidden, also those that may be made remunerative ... that consists of gold, bullion, silver, and of plate ware ...**
>
>

996-1
August 14, 1926

This reading was correct in saying that "oil production [would be] very low" in the Bimini area. Years later, this was proven decisively with the drilling of the Great Isaac Island 1 exploratory oil well (see Figs. 23 and 24) just north of Bimini. The wildcat encountered no exploitable oil deposits.

The treasure hunt, begun in 1926, encountered difficulties. And so, in February 1927, Cayce went with the treasure hunters to Bimini for three days. There he gave four readings while lying on the ground in the vicinity of the sought-after treasure. But nothing could be located. After returning to Miami, Cayce gave another reading in which he was asked " . . . why were we unable to locate this [treasure?]" The answer:

> **Yes, we have the information as has been given regarding treasure buried on Bimini Island, with those who made diligent search for same without results . . . Now we find that, not because of the information being incorrect . . . [but because] . . . the trouble lies within that of the one [Edgar Cayce] through whom information is given; for these sources from which the information comes to the material world are from a universal and infinite source, but the channel of same is of the carnal or material plane. Hence we know sin lies at the door, and in that information as has been given respecting same, that the house must be set in order.**
>
> **996-8**
> **February 7, 1927**

There are clear implications in several readings that there was something wrong or insufficient about the information received and about the manner in which it was acted upon by the recipients, that was preventing them from reaching their materialistic objectives. For example, in 996-10 we read that the psychic information on buried treasure on South Bimini was:

> ... given in such a way and manner that those carrying out same did not succeed in locating same ... [and that] ... information as is given through the manifestation of the Universal Forces as are manifested through ... Edgar Cayce ... is for the uplifting, and should never be used for other than that ...

Cayce was asked about yet another money-making scheme, this one for financing and developing a resort city on the Bimini Islands. The answer was very encouraging:

> First we would give, this would not be near as large an undertaking as it appears on the face of conditions for ... There will be found many, many, many sources of revenue for those undertaking such a project, for these mountaintops—especially that along the north and eastern shores of the north and northern portion of the south island— will produce many various minerals, and various other conditions that will be remunerative when the projects are undertaken; and well that the ones that do such labors—as the dredging as necessary ... be followed close in their operations, for these will uncover many various conditions that may be turned into dollars—and dollars—and dollars!
>
>   996-12
>   March 2, 1927

The reading also explained how the reclamation of land should be undertaken and how financing might be obtained, concluding with:

This [inlet between north and south Bimini] also lends the modes of manufacture—of electro-hydro power in the waves, if necessary, by the tides—such as have been and are being builded in the Bay of Fundy—for with the walls as may be built in the western coasts of the inlet—which may be closed or left open or builded for the purpose of an inland sea for the boat, the port, and for the fishing, bathing, and the like—this may be builded in such a manner on the northern shore of the south, and the southern shore of the north island, as to facilitate the power sufficient to electrify the whole of the lands that may be acclaimed and reclaimed.

The walls just mentioned would be located across the inlet at Entrance Point (Fig. 25). Strong tidal currents in this inlet lend credence to the suggestion that the inlet be dammed to make a hydroelectric plant. Currents range up to four knots at times of spring tides, and there is only a short period of slack water at times of tide change. In addition, there is an asymmetry to the tidal-current curve such that the ebb flow, to the southeast out of the lagoon, lasts approximately twice as long as the flood flow. This situation is ideally suited for generating hydroelectric power.

As for Poseidia, Cayce said:

A temple of the Poseidians was in a portion of this land.

996-12
March 2, 1927

This temple was mentioned in another reading years later, given for a woman who was told that she had once incarnated in Atlantis:

Before that the entity was in the Atlantean land during the period of the second breaking up, when the islands—or Poseidia—became the main portion of the activities . . . In that experience the entity was among the children of the Law of One who accepted and forsook much of those activities because of its close associations and companionships with one of the sons of Belial. This brought consternation to the entity, and also those influences the application or use of which brought destruction to the land. And Poseidia will be among the first portions of Atlantis to rise again. Expect it in sixty-eight and sixty-nine; not so far away!

958-3
June 28, 1940

There is no evidence that portions of a sunken Atlantean continent rose above the ocean surface near the Sargasso Sea (Poseidia's approximate location) in 1968 and 1969. Nor is there any evidence for rising in the vicinity of Bimini. We note, however, that the reading only says "rise," not emerge. Also, no reading ever said specifically that the remains of the Poseidian temple near Bimini would be rising again. However, the following reading about a prehistoric well on Bimini could be interpreted to mean that the submerged Poseidian temple in the vicinity of the islands could be expected to rise with the start of the major Earth changes (presumably in 1998).

Nearly 10 years after Cayce's visit to Bimini a woman told him that she had flown to the islands and found a freshwater well marked or walled around the top with stones of peculiar composition and strange symbols. So she got a reading in which she asked:

Q Could the well in Bimini be promoted and re-

constructed? [See Fig. 25 for the author's best guess as to the location of this well.]

A There has been much given through this source as to how that particular portion of what was the Atlantean period might be developed. While it would make for much outlay in money, as ordinarily termed, there are certain interests that would join in such an undertaking. As those of the Dodge interests, as given. For it could be established as a center for two particular purposes; a regeneration for those with certain types of individual ailments (not only from the well, or water from same, but from the surrounding waters—because of the life in same), and a center for archaeological research. And as such activities are begun, there will be found much more gold in the lands under the sea than there is in the world circulation today!

As to how, this should be considered seriously from many varied angles that exist. For, as understood, there are those conditions as related to the varied powers that are in power or in affluence as respecting activities of any nature there. And as they exist in the present there are some complications for agreements, contracts, the lettings of this, that or the other.

But this should not be left alone; it should be considered from many angles.

Also aid may be induced from the varied societies that have been formed for the study of geological and archaeological activities, or such. For much will be found.

And, as may be known, when the changes begin, these portions will rise among the first.

587-4
July 1, 1935

What *is* clearly stated about Atlantis rising is given in the following reading where one reads that Atlantis is "rising again."

> **It would be well if this entity were to seek either of the three phases of the ways and means in which those records of the activities of individuals were preserved—the one in the Atlantean land, that sank, which will rise *and is rising again;* another in the place of the records that leadeth from the Sphinx to the hall of records, in the Egyptian land; and another in the Aryan or Yucatan land, where the temple there is overshadowing same. [Emphasis added.]**

2012-1
September 25, 1939

What was going on at Bimini during 1968 and 1969 that might relate to the finding of a "temple of the Poseidians" at that time? From 1965 to 1968 I conducted intermittent field investigations at Bimini, including a search for a vein of "gold, spar, and icthyolite" that the readings on Bimini said might be found there. I realized that it might be necessary to penetrate the sediments and limestone crusts in order to determine the nature of the rocks that composed the ground surface which has lain buried some 11,000 years. A preliminary reconnaissance in that year revealed that minor dredging near East Swell Inlet (also known as East Well or Muddy Creek Inlet) had not turned up anything significant in this part of the "north . . . portion . . . of the north island." Nor had any water wells, trenches, or canals been dug sufficiently deep or in the right places to permit encounter with the supposed pre-11,000-year-old rocks. In 1965, however, a channel was dredged for the Buccaneer Point Marina at the northern tip of South Bimini.

While inspecting carbonate sand and limestone dredge spoil (Fig. 25, point A) from the deepest part of the entrance channel, I found a pebble of serpentinite, completely encrusted with limestone. An expert petrographer examined a thin section of the pebble and found peridotite, chromite, bastite after hypersthene, kink-banding deformation in the pyroxene, and other features characteristic of serpentinized periodites. The evidence strongly argues against the pebble being a piece of ship's ballast.

I also found at this site a small, limestone-encrusted black pebble. Dissolving away most of its limy encrustation revealed the stone to be a jet black angular fragment, about 15-mm in its longest dimension. Laboratory analysis showed it to be of extremely high carbon content (98.3 %), even for high-rank anthracite coal, and perhaps a substance that may not be coal at all. We will return shortly to these two pebbles.

I began to wonder what geological evidence there might be that the Bimini Islands are former "mountaintops," as mentioned in reading 996-12. Was this merely a reference to the appearance of the Bimini portion of the Great Bahama Bank some 20,000 years ago when global sea level was about 425 feet lower than at present? Or had the Bimini Islands area been in a tectonically upthrust position during the Atlantean period? I thought that if I could take relatively large-diameter rock cores to a great enough depth, it might be possible to find some evidence of the ancient land surface common to such supposed mountaintops.

While conducting a 1966 field project funded by the Edgar Cayce Foundation, I was able to hire a drilling barge that happened to be operating near Bimini. We put down four, 4-inch diameter drill holes just offshore of South

Bimini. (Drilling locations are shown on Figure 25.) The borings reached the following depths, in feet, below mean low water (MLW): F (-40), G (-45), H (-40), and I (-45). Examination of the drill cuttings and discontinuous core samples revealed carbonate material, exclusively. A study of the ratios of low-magnesium calcite to aragonite in the carbonate rock core samples revealed the absence of aragonite in all samples below 26 feet MLW.

This type of information indicates that a former subaerial (land) surface lies about 26 feet below MLW in this area of the borings. If so, the "mountaintops" of the readings are probably composed entirely of carbonate rock. They were submerged and then covered by recent carbonate sediments during rising sea level over the last several thousand years.

An alternate view builds on the picture that geoscientists have worked out for the northwest Bahamas, in which the Banks were built by coral reefs on a spreading ocean crust with volcanic extrusions. This took place as Africa split off from Florida, during the opening of the Atlantic some 150 to 180 million years ago. One such extrusion might have produced a twin-peaked volcanic mountain that became surrounded and then covered by coral reefs as the bank slowly subsided over geologic time. This could explain the puzzle of the pronounced westward projection of the northwestern Great Bahama Bank into the Florida Straits (Fig. 24).

If such a hypothetical twin-peaked volcano underlies North and South Bimini it would probably do so at considerable depth beneath the carbonate rock that makes up the immediate subsurface. Robustly expressed volcano landforms typically undergo considerable erosion prior to their final submergence and entombment by reef limestones in areas of slowly subsiding crust like the Bahamas. Also, during the 18,000 years after the end of the last glacial period, sea level rose so rapidly that the reefs did not develop with sufficient rapidity to maintain the subaerial expression of most Bahamian islands.

If a volcano does lie deep beneath the Bimini Islands, we might have an answer to one of the more puzzling aspects of the readings on Bimini. The first has to do with the claim in 996-12 that a northeast-southwest oriented, workable vein of "gold, spar, and icthyolite" could be found during construction of a wall between North and South Bimini. This vein would lie at "the twelve- to fifteen-foot levels" (presumably beneath the bottom of the Bimini Inlet). It is my conjecture that the vein minerals would be found in a fracture in the carbonate country rock that resulted from an episode of injection of fluids from the volcano below. Spar minerals could have resulted from such activity and gold could be carried in as well. Gold has been found in many places in volcanic terranes of the ocean floor. But proponents of the theory of sea-floor spreading would not agree with this scenario because the volcanic edifice beneath the Bahamas is believed to be too cold now to allow for renewed volcanism within the last few million years. Only if the North Atlantic Ocean basin underwent volcanic upheavals during those phases of destruction of Atlantis at 19,400, and between 12,657 and 11,285 years ago, might one be able to assume that short-lived volcanic activity, sufficient to produce the vein minerals, might have taken place beneath Bimini.

As for the "icthyolite" that the reading also mentioned could be found at Bimini, there are no known "icthyolite" minerals in the literature of mineralogy. Perhaps Cayce's stenographer wrote the wrong word and minerals associated with *ijolite* rock was meant. Interestingly, these rocks owe their characteristics to contamination of magma by limestone and may be found in serpentinized areas. Ijolites are highly alkaline, low silica, aegirine-nepheline rocks that contain a long list of accessory minerals such as apatite, sphene, calcite, melanite, phlogopite, sodalite, perovskite, wallastonite, cancrinite, pectolite, and zeolites—any of which may increase sufficiently in amount to become a major constituent. Some of the minerals just named are "spar" minerals.

Whether there is a linear deposit of gold and spar artifacts lying upon the carbonate rock of a buried and submerged mountaintop, or whether the gold and spar minerals would reside in a mineralized fracture of the former mountaintop is open to speculation.

In 1967 I found several fragments of tough gray and brown slate in dredge spoil on both sides of the inlet. It may be ship's ballast, although if it is, it's a type of ballast not commonly found in the Bahamas.

In 1969, the explorer J.M. Valentine described what he called an "archaeological enigma, consisting of 'pavement-like' stones at 15 feet off North Bimini" (off Paradise Point on Fig. 25) on the Gulf Stream side of the islands. This sparked speculation that the stones were a submerged seawall or roadbed built by Atlanteans. Sections of "pillars carved from natural stone" were also said[9] to lie at shallow depths off Entrance Point (Fig. 25), "exactly where and when Edgar Cayce prophesied the re-emergence of Atlantis."

In 1970, these claims were investigated by two marine geologists and a marine biologist.[10] They concluded that the stones were natural limestone, and the "pillars" were composed of cement, manufactured around 1800, that had hardened in barrels and probably been lost overboard in a shipwreck. Two pieces of a fluted marble column were also found with the cement cylinders and were presumed to be associated with the cement as part of a load of building materials being transported by a ship that ran aground on the ocean side of Entrance Point.

In 1980, M. McKusick, an anthropologist at the University of Iowa, and E. Shinn, a geologist with the U.S. Geological Survey in Miami Beach, reinvestigated the so-called "Bimini road" and stated that "although under 15 feet of water, the [submerged stones] are beachrock of recent geological origin."[11] They obtained seven carbon-14 dates from samples of the submerged beachrock. The dates range in age from 2,745 to 3,510 years before the present. They concluded that based on the rate of sea-level rise over the past 5,000 years, about six to seven feet of the 15 feet of sea water observed over the beachrock can be accounted for by worldwide sea-level rise over the last 5,000 years and that "the remaining eight to nine feet of sea may be explained by the undermining of sand, allowing the beachrock to gradually settle, an erosive process that is seen in various stages in many parts of the Caribbean."

Herein lies cautionary advice for anyone interested in investigating the Cayce readings on Atlantis at Bimini. Avoid accepting as evidence for Atlantis natural surficial phenomena that make provocative patterns or which are too young in radiocarbon age to be associated with a civilization claimed to have gone under the waves 11,900 years ago. Douglas G. Richards gives an example of such objectivity in his work dealing with the interpretation of patterns in the sea grass and the submerged beachrock in the shallow waters around Bimini *(Venture Inward*, Mar./Apr. 1986). He agrees with earlier scientists' conclusions that the "Bimini road" is nothing more than slabs of submerged beachrock.

Returning now to the problem of locating the Poseidian temple and to explaining the two pebbles that I recovered from deep dredged material, reading 440-5 says that records of the construction of an Atlantean power station may be found in the *sunken* portion of Poseidia where "a portion of the [Atlantean] temples may yet be discovered, under the slime of ages of sea water—near what is known as Bimini, off the coast of Florida" and that the center of the original power station was lined with "a something *akin* to asbestos, with the combined forces of bakelite or other nonconductors . . . " [Emphasis added.]

Q Give an account of the electrical and mechanical knowledge of the entity as Asal-Sine in Atlantis.

A Yes, we have the entity's activities during that experience. As indicated, the entity was associated with those that dealt with the mechanical appliances and their application during the experience. And, as we find, it was a period when there was much that has not even been thought of as yet in the present experiences.

About the firestone that was in the experience did the activities of the entity then make those applications that dealt with both the constructive and destructive forces in the period.

It would be well that there be given something of a description of this . . . as to how both constructive and destructive forces were generated by the activity of this stone.

In the center of a building, that today would be said to have been lined with nonconductive metals, or nonconductive stone—something akin to asbestos, with the combined forces of bakerite [bakelite?] or other nonconductors that are now being manufactured in England under a name that is known well to many of those that deal in such things.

The building above the stone was oval, or a dome wherein there could be or was the rolling back, so that the activity of the stone was received from the sun's rays, or from the stars; the concentrating of the energies that emanate from bodies that are on fire themselves—with the elements that are found and that are not found in the Earth's atmosphere.

The concentration through the prisms or glass, as would be called in the present, was in such a manner that it acted upon the instruments that were connected with the various modes of travel, through induction methods—that made much the character of control as the remote control through radio vibrations or directions would be in the present day; though the manner of the force that was impelled from the stone acted upon the motivating forces in the crafts themselves.

There was the preparation so that when the dome was rolled back there might be little or no hindrance in the application ... we find it was a large cylindrical glass (as would be termed today), cut with facets in such a manner that the capstone on top of same made for the centralizing of the power or force that concentrated between the end of the cylinder and the capstone itself.

As indicated, the records of the manners of construction of same are in three places in the Earth, as it stands today: In the sunken portions of Atlantis, or Poseidia, where a portion of the temples may yet be discovered, under the slime of ages of sea water—near what is known as Bimini, off the coast of Florida. And in the temple records that were in Egypt ... [and so on].

440-5
December 19, 1933

This reading *may* have the answer to the location of the Atlantean temple of records. If it is the same as the temple holding the record of the method of constructing the cylindrical crystal, then it is located near Bimini and not in the Atlantic near the Sargasso Sea.

As for searching for the temple "under the slime of ages of sea water," such a sea-floor environment is more likely at a depth of 450 fathoms (2,700 feet) than at 50 fathoms (300 feet). The *generalized* near-surface, sediment-type map for the northern Bahamas shown by Sheridan and others (Fig. 23, p. 164 in footnote 6), indicates that sands make up the ocean floor between the Bimini shoreline and the area of rough topography of Figure 24. The area of rough topography consists of largely carbonate sands shaped into contourites, or features formed by contour-following bottom currents. In still deeper water, there are deep-water mounds of limestone, formed apparently by submarine lithification of carbonate mud, sand, and skeletal debris. Beyond about 15 km (9.3 miles) offshore to the west of Bimini, the bottom is composed of ooze, the closest material to the term "slime" in the reading. In marine geology, ooze is generally a pelagic sediment, consisting of at least 30 percent skeletal remains of calcareous or siliceous pelagic organisms, and clay minerals. *Slime,* in *Webster's Dictionary,* is "viscous mud." It's possible that patches of slime occur closer to Bimini than 9.3 miles offshore to the west.

Shipboard studies of the sea floor with side-scanning sonar and a geographic positioning system (GPS) for ship location, followed up by underwater photography of likely targets, could facilitate the search for the temple remains.

But what of the black pebble that I found and Cayce's speculation that something *akin* to bakelite was used to line the building containing the Atlanteans' firestone? Bakelite is also black, contains considerable carbon, and is a well-known electrical insulator. It was being manufactured in England at the time of Cayce's reading in 1933. It seems worth musing upon whether this angular pebble is either one of the highest rank coal specimens ever found or a piece of the lining of an Atlantean power station that was blown to pieces many thousands of years ago in an enormous man-made, volcanolike eruption.

Recall this reading:

> ... [men on Atlantis] **brought in the destructive forces as used for the peoples that were to be the rule, that combined with those natural resources of the gases, of the electrical forces, made in nature and natural form the first of the eruptions that awoke from the depth of the slow cooling Earth, and that portion [of Atlantis] now near what would be termed the Sargasso Sea first went into the depths.**
>
> **364-4**
> **February 16, 1932**

The piece of serpentinized peridotite that I found is clearly an exotic rock type for the carbonate-sediment environment of Bimini. Interestingly, the U.S. Ocean Drilling Program recently cored serpentinized peridotites from the western wall of the submerged mid-Atlantic Ridge.[12] These rock cores were recovered about 2,000 miles due east of Bimini, from an area of the sea floor beneath the Sargasso Sea.

As mentioned earlier, the very destructive event described in the reading immediately above probably occurred 19,400 years ago (364-11). Two other readings indicate that there was a sort of repeat of this type of destruction between 12,700 and 11,900 B.P.

Are the peridotite and black-stone fragments natural and anthropogenic materials, respectively, from Atlantis? The honest skeptic says, you are just piling speculation upon conjecture here. Even granting that Atlantis might have existed, how can you possibly suggest that the two small fragments recovered from deep within the sands at Bimini Inlet could reflect an enormous explosion that occurred in the mid-Atlantic 10,000 to 20,000 years ago and 2,000 miles to the east of Bimini?

I'll answer this question with the following contemporaneous account. Then decide, by comparison, how improbable my story might be.

Geologists believe that an asteroid blasted the Earth at the end of the Cretaceous, around *65 million years ago,* in what is now the Yucatan Peninsula. This catastrophic event of *extraterrestrial origin* caused the mass extinction of many species at the end of the dinosaur age. At the March 1996 meeting of the Lunar and Planetary Science Society, F. Kyte, a UCLA geochemist, showed slides of a chip of rock that he thinks may be a tiny fragment of a *10-kilometer* asteroid that is believed to have struck Earth. "From the age and makeup of the *3-millimeter chip* found in the ooze *at the bottom of the North Pacific [over 4,000 miles away],* Kyte is 'personally convinced that it is a piece of the bolide,' a conclusion other researchers say is at least plausible." (R.A. Kerr, *Science,* 29 March 1996, p. 1806. I've used italics to draw attention to features of the report in *Science* that relates to similar, but less extreme, aspects of my own conjecture.)

These musings of mine may be either dreamy abstraction or productive inspiration. Choosing the latter option, here is a modest proposal for research to verify, if possible, the supposed remains of Atlantis at Bimini. Obtain permission from the Ministry of Lands and Works,[13] Government of the Bahamas, Nassau, to conduct exploratory dredging "along the north and eastern shores of the north and [along the] northern portion of the south island" (see reading 996-12 quoted earlier). Then hire a firm with an hydraulic dredge properly sized to remove sand to the 12,000-year-old depth in the carbonate sediment or to the limestone rock surface of "the former mountaintop." Follow the dredging operation moment by moment and carefully examine the contents of the dredged material for natural minerals and rocks or man-made substances exotic to the surrounding carbonate environment.

If the dredging efforts are successful in turning up ancient artifacts or mineral fragments, they may ultimately fulfill the 996-series readings on the projected development at Bimini, as well as the associated discovery for mankind of remains of "lost Atlantis."

CHANGES IN THE OCEAN FLOORS

If portions of sunken Atlantis in the vicinity of the Sargasso Sea are now *rising*, they are doing so at a slow enough pace to prevent upsetting the ongoing worldwide rise in sea level due primarily to (1) melting of glacial ice and (2) to the thermal expansion of ocean waters as they respond to worldwide climate warming. What will happen to sea level if the rate of uplift of Atlantis suddenly changes? And what will be the effect on ocean-basin capacities and sea level of other lands rising as well, as predicted in 3976-15 and in 1152-11?

A . . . There will be new lands seen off the Caribbean sea, and *dry* land will appear . . . and in the

Antarctic off of Tierra del Fuego *land,* and a strait
with rushing waters.

3976-15
January 19, 1934

In the next few years lands will appear in the Atlantic as well as in the Pacific. And what is the coast
line now of many a land will be the bed of the
ocean. Even many of the battlefields of the present
[1941] will be ocean, will be the seas, the bays, the
lands over which the *new* order will carry on their
trade as one with another.

1152-11
August 13, 1941

The Antarctic Peninsula lies directly south of Tierra del
Fuego at the tip of South America. This peninsula is one
of the discrete crustal blocks that comprise West Antarctica. It is an Andean-type magmatic arc that resulted
from subduction of Pacific Ocean lithosphere beneath
its western margin. "Trends of both fracture zones and
magmatic lineaments strongly support a link between
faulting in the Antarctic Peninsula magmatic arc and offshore tectonics."[14] The upheavals in the Antarctic mentioned in the readings could have the ability to affect the
"offshore tectonics" just mentioned in such a way as to
produce land "in the Antarctic off of Tierra del Fuego."
And upheavals there could be sufficient to cause a "strait
with rushing waters," between the Antarctic Peninsula
and the tip of South America, or somewhere in between.

As for geophysical evidence that might support new
lands appearing off the Caribbean or in the Atlantic and
Pacific Oceans, some scientists interested in global tectonics are beginning to talk about "oceanic plateaus,"
"large igneous provinces," and "mid-Pacific super

swells" that reflect more than the effects of just one, relatively small-diameter rising mantle plume. But these concepts are still in a formative stage.

The phrase, "In the next few years . . . " used in 1941 in reading 1152-11, seems to be quite wrong, unless the psychic source is using figurative language, as "in only a relatively short while." Then again, the phrase in question could be read as an expansion on the sequence of events given in the sentence previous to it. There we read, " . . . many portions of the east coast will be disturbed, as well as many portions of the west coast, as well as the central portion of the U.S." Thus, "in the next few years" could mean the years *following* the disturbances just listed. And those disturbances are scheduled to begin in 1998.

Finally, here's a reading, hard to understand, that tries to explain to our finite minds why we have no solid records of Atlantis or other lost continents. And it carries a telling spiritual message as well.

> **Then why, as has been said, is God mindful of an individual soul?**
>
> *Spirit!* **For our spirit, that is a portion of His Spirit, ever bears witness with His Spirit as to whether we be the children of God or not.**
>
> **Then through Mu, Oz, Atlantis—with the breaking up of these—why, *why* no records of these if there were the civilizations that are ordinarily accredited to them by the interpreting of the records made by entities or souls upon the skein of what? Time and space?**
>
> **But he only that has recognized patience within self may indeed make the record as an experience in the consciousness of any.**
>
> **Each individual spirit then is only a portion of**

His Spirit. Not that God is separated. But in His love, in what we call infinite love, boundless, the unbounding grace and mercy and patience and love and long-suffering, these have brought to the Father the thought of the lack of, the wonderment of, companionship.

262-115
October 24, 1937

8

Danger Lands/ Safety Lands

The Cayce readings predict that this is what may happen globally: Starting in 1998, the poles of the Earth may begin to shift. The primary causative mechanism for this polar movement occurred in 1936 when some portion of the interior of the Earth moved with respect to Earth's axis of rotation. Coincident with this internal movement, plumes of hot mantle material began to rise toward the Earth's surface. The initial effects of these rising plumes has been expressed in volcanic activity during recent decades.

The closer we get to 2000 the more obvious will be these upheavals of mantle material in various parts of the planet. Rising plumes will produce accelerating vertical and horizontal geologic stresses on the Earth's crust. These stresses will coincide with other mantle movements required to accommodate the movement of the equator as it adjusts to changing density configurations in the Earth's outer shell. Upheavals of the crust and mantle will occur in the Arctic and Antarctic, causing eruption of volcanoes in the torrid areas. Some upheavals may reflect melting of glacial ice, especially in West Antarctica. All such upheavals will further affect the density structure of the Earth's outer mantle and crust, contributing to the need for the equator to move and for the poles to shift.

Mantle shifts related to the moving equatorial bulge will cause further volcanic eruptions in the torrid areas. The Earth's lithospheric plates will experience accelerated movements, both vertically and horizontally. The ocean floor will rise in parts of the Atlantic and the Pacific. The northern portion of Japan will be submerged, as will southern portions of South Carolina and Georgia. Portions of the lost continent of Atlantis, including the ruins of an important Atlantean temple, will rise above the ocean surface. Land will appear off the Caribbean Sea, presumably in the Atlantic east of the Caribbean, and off the southern tip of South America. Severe shaking will occur in South America, and seismotectonic events will take place in nearly all of the western United States. Significant disturbances will occur in parts of the central U. S., as well as portions of the eastern states. The entire world will witness many and varied geophysical disturbances. Northern Europe will be changed over a short period—perhaps only a decade—either by uplift of the crust or by the sudden warming of the climate there.

In the Western Hemisphere, the early stages of the accelerating changes in the Earth's surface will see the breakup of terranes along the U.S. west coast. Submergence of the coast of South Carolina and Georgia will also occur early in the period. Most of Los Angeles and San Francisco will be destroyed. The greatest physical change in America will be to Connecticut and southern New York, including New York City.

Precise times for the seismotectonic events cannot be given because locations and dates for Earth changes are implemented at the discretion of the Universal Forces. There is one exception to this mystery about timing. Significant eruptions of either Mt. Pelée or Mt. Vesuvius will signal that, sometime during the following three months, massive earthquakes will strike coastal Southern California and the areas between Salt Lake and southern Nevada. More widespread and intense shaking will occur in the Southern Hemisphere, in South America, Antarctica, Indonesia, and probably New Zealand.

The bulk of the Earth changes will occur because Nature—planet Earth—has her preordained cycle of development, in keeping with a divine plan. As the Earth moves into a new phase—marked by the transition from the Piscean to the Aquarian age in 1998—certain crustal and other readjustments become necessary. The timing and severity of some of the inevitable Earth changes may be delayed or mitigated by those individuals and groups who have not forgotten God and who seek to show the love of the Father to their faltering, erring brothers—"but to those that seek, not those that condemn." (3976-15)

The remainder of the potential Earth changes may also depend upon the behavior of souls who have been particularly anti-humanity (anti-God) during the past Age. It is still not too late for such individuals to "right about face." But the choice is theirs alone, as it was for the person mentioned in the following reading:

Before that the entity was in the Atlantean land, when there were those activities in which there was the destruction of the isles.

The entity chose to remain, and was destroyed with those upheavals.

3020-1
May 25, 1943

That person's choice was based on free will.

Man alone is given that birthright of free will. He alone may defy his God!

5757-1
June 24, 1940

People living in one of the areas designated in the readings to be affected by Earth changes may ask: What should I do now? Are there any safe places to live? Should I move to a safety land?

Answers to these questions will be different for each of us. Through proper mental preparation, improved attitudes, good works, and prayers and meditation, we're promised that we'll obtain the insight needed to help us know what to do. The orientation we need in this pursuit seems to be well put in the following Cayce readings:

Q In what way have I failed to heed the advice of my Maker, in my actions or deeds, or wherein have I been lacking?
A Study to show self approved. Turn within self. No man has the right to find fault with his brother. Neither do the angels that stand before the throne of mercy find fault, but rather present that which the individual soul has done with his knowledge, his intellect, his understanding. For, thus does the

soul find the relationships to the Maker; whether there is the god of light and love and hope, or that which was separated from same that makes for despair, for night, and for those things that do hinder the approach.

Look, then, into self. Answer to self. For, each individual must so live each day that he may look into the face of that he has spoken, that he has lived, and say: "By this I stand to be judged before myself, before my God."

257-123
September 29, 1933

Do that which is good, for there has been given in the consciousness of all the fruits of the spirit: fellowship, kindness, gentleness, patience, long-suffering, love; these be the fruits of the spirit. Against such there is no law.

Doubt, fear, avarice, greed, selfishness, self-will, these are the fruits of the evil forces. Against such there *is* a law. Self-preservation, then, should be in the fruits of the spirit, as ye seek through any channel to know more of the path from life—from good to good—to life; from death unto life, from evil unto good. Seek and ye shall find. Meditate on the fruits of the Spirit in the inner secrets of the consciousness, and the cells in the body become aware of the awakening of the life in their activity through the body. In the mind, the cells of the mind become aware of the life in the spirit. The spirit of life maketh not afraid.

5752-3
November 8, 1933

The entity was an Atlantean, as indicated. This

experience was indicated because there had been the opportunity in self, in which there was the use of spiritual ideals for self-indulgence. Hence such entities, in each period of activity in the material plane, will either make reparations or—rather should it be said—turn to the law of grace instead of the law of karma, and will come out of the experience as a wonderful success or a miserable failure. They are extremes, just as the entity itself has found in the present. There are periods when there is the decision that something is desirable and you don't stop until you have it! These are extremes. Time and space and patience are most needed oft, that few souls or individuals are willing to pay the price for—until they grow to be such, you see ...

Q In view of the uncertainty of existing conditions, did I act wisely in establishing my home in Norfolk [Virginia]?

A It's a mighty good place, and a safe place when turmoils are to rise; though it may appear that it may be in the line of those areas to rise, while many a higher land will sink. This is a good area to stick to.

Q Where should I buy a home?

A Anywhere in the area of what may be called the Tidewater section.

2746-2
November 11, 1943

Mark A. Thurston, executive director of the A.R.E. and a longtime student and interpreter of the Edgar Cayce readings, provides an interesting perspective on the readings' concept of "safety lands" in his book, *Visions and Prophecies for a New Age:*

It would be inconsistent with other Cayce readings to think that a soul is safe just because physical harm is unlikely. The entire thrust of the philosophy of the readings is that real security is an inward thing. We are truly safe when we are right with God, when we are living in accordance with the spiritual ideal we have set. For example, in 1942 one person asked in a reading if he was safe in his home town in New York—safe from bombing or enemy attack. Cayce's answer: "Why should he not? if he lives right!" (257-239, January 15, 1942)

What is needed is another hypothesis concerning the meaning of Cayce's "safety lands" concept. Not a theory which contradicts the possible literal interpretation, but instead a complementary one which suggests multiple levels of meaning of that concept. A promising approach is that cities, regions, or even states can create, by the attitudes and actions of their people, a kind of group consciousness and aura which gives a location a particular vibration. That vibration, if it is of spiritual attunement, can afford to its people a sort of protection . . .

This notion of group vibration for a location is found in at least one reading which says: "Each state, country, or town makes its own vibrations by or through the activities of those that comprise same; hence creates for itself a realm in which the activities of each city, town, state . . . may be in the realm of those forces where the activities bring the associations through relativity of influence in the material plane." (262-66, July 11, 1934)

. . . Even though some of these places [described in a 1941 reading] may be physically shaken—even though every one of them will feel the effects in some way if the Earth changes take place—they may, nevertheless, be places of special opportunity and of psychological and spiritual safety in times of change. What is exciting is that we do not have to move to one of those existing "safe" locations. We can each work to build that status for the community in which we already live.[1]

Finally, it seems that our true safety in times of Earth changes will somehow be related to the movement of the Spirit over the Earth, magnifying itself in the hearts, minds, and souls of humankind, in the new age opening before us. Consider this reading, given in the depths of the Great Depression, but as applicable today as it was then:

With the present conditions, then, that exist— these have all come to that place in the development of the human family where there must be a reckoning, a one point upon which all may agree, that out of all of this turmoil that has arisen from the social life, racial differences, the outlook upon the relationship of man to the Creative Forces or his God, and his relationships one with another, must come to some *common* basis upon which all *may* agree. You say at once, such a thing is impractical, impossible! What has caused the present conditions, not alone at home but abroad? It is that realization that was asked some thousands of years ago, "Where *is* thy brother? His blood *cries* to me from the ground!" and the other portion of the world has answered, *is* answering, "Am I my brother's keeper?" The world, *as* a world that makes for the disruption, for the discontent—has lost its ideal. Man may not have the same *idea.* Man—*all* men—may have the same *Ideal!*

As the Spirit of God once moved to bring peace and harmony out of chaos, so must the Spirit move over the Earth and magnify itself in the hearts, minds and souls of men to bring peace, harmony and understanding, that they may dwell together in a way that will bring that peace, that harmony, that can only come with all having the one Ideal; not the one idea, but "Thou shalt love the Lord thy God with all thine heart, thy neighbor *as* thyself!" This [is] the whole law, this [is] the whole answer to the world, to each and every soul. That is the answer to the world conditions as they exist today.

How shall this be brought about? As [they] each in their own respective sphere put into action that they know to be the fulfilling of that as has been from the beginning, so does the little leaven leaven the whole lump.

Man's answer to everything has been *power*— power of money, power of position, power of wealth, power of this, that or the other. This has *never* been *God's* way, will never be God's way. Rather little by little, line upon line, here a little, there a little, each thinking rather of the other fellow, as that that has kept the world in the various ways of being intact—where there were ten, even, many a city, many a nation, has been kept from destruction. Though ye may look upon, or feel that that which was given to Abram—as he viewed the cities of the plain and pled for the saving of same— was an allegorical story, a beautiful tale to be told children—that it might bring fear into the hearts of those that would have their *own* way—may it not come into the hearts of those now, today, wilt *thou,* thine self, make of thine *own* heart an understanding that thou must answer for thine own brother, for thine own neighbor! and who is thine neighbor? He that lives next door, or he that lives on the other side of the world? He, rather, that is in *need* of understanding! He who has faltered; he who has fallen even by the way. *He* is thine neighbor, and thou must answer for him!

3976-8
January 15, 1932

Specific safety lands given for North America in the Cayce readings were listed as portions of Ohio, Indiana, and Illinois and much of the southern and eastern portions of Canada, as well as the Norfolk and Virginia Beach, Virginia, areas (readings 1152-11 and 2746-2). Safety lands in other countries were never specified, but certain areas were mentioned as destined for seismotectonic disturbances and others may be inferred as candidates for such disturbances. Take South America, for example. Figure 20 shows that the highest risk areas lie along the chain of mountains adjacent to the Pacific Coast. The eastern section of this vast continent is considered relatively safe. The same may be said of the western portion of southern Africa and perhaps all of Australia. These appear to be the safest places in the Southern Hemisphere.

Another indicator of safe regions may be taken from forecasts of agriculture production. Reading 3651-1 advised a Cayce client to get involved in the granary business in the wheat country of Montana. The reading added:

Learn to deal, then, with those in portions of Saskatchewan as well as in the pampas area of the

Argentine, as well as in portions of South Africa. For these rich areas, with some portions of Montana and Nevada, must feed the world!

3651-1
January 15, 1944

Saskatchewan is currently a significant grain producer and lies in the "southern portion of Canada," mentioned in reading 1152-11 as to be among the safety lands of North America. Argentina's richest natural resource is the fertile plain known as the pampas, which fans out almost 500 miles south and west of Buenos Aires to south-central Argentina. The pampas produce large quantities of wheat, corn, soybeans, and sunflower seeds. It fits into our geophysical interpretation (see Chapter Six on South America) in which *eastern* South America will be amongst the safer areas of that continent, during those times when it is "shaken from the uppermost portion to the end" during the Earth changes.

In 1968-1971, South Africa's average wheat production amounted to 2.5% of world production, Montana about 0.3%, and Nevada only slightly above 0.0%.

What is interesting about these statistics and the prognostications of the above reading are the countries left out of the picture, namely, countries of the former Soviet Union (together producing 28% of world wheat production in 1968-1971), China (9%), Europe (22%), Australia (3%), India (4%), Turkey (3.3%), other Asian countries (7%), and the United States, other than Montana (about 12.3%). We can only assume that far more than seismo-tectonic events associated with pole shift will be involved in such a drastic diminishing and redistribution of world grain production after 1998.

Changes in world climate due to countries' new alignments with the world climate belts and, probably, changes in mean global temperature will play a strong role in such predicted grain-production changes. Significant eruptions of Torrid Zone volcanoes will contribute to global temperature changes.

Following the pole shift, we can expect changed interactions between the new locations of the present grain-producing areas and Earth's climatic belts.

This may explain much of the loss of grain production for the current producing countries as described above. Where rainfall amounts are lessened, there will be great demand for fresh water for irrigation of crops. Parts of Ohio, Indiana, and Illinois may be impacted favorably in this regard if the waters of the Great Lakes are to empty into the Gulf of Mexico (1152-11). Farmland in the northern sections of these states and along waterways leading to the Gulf will have available adequate irrigation water even if the climate becomes drier there.

An example of what can happen with even a slight loss of adequate rainfall in a grain-producing area is provided by the tenuous climatic situation in the U. S. Great Plains, the breadbasket of America. This has been highlighted in an article by William Stevens, "Great Plains or Great Desert?" (*New York Times*, 5/28/96), in which he says that only ten years of drought could re-create a Sahara in the plains states.

"Vast stretches of the plains, from Texas and New Mexico to Nebraska and South Dakota, consist of classic sand dunes covered thinly by vegetation . . . When atmospheric circulation settles into a characteristic pattern of no rain and high heat over the region, the vegetation dies. The dunes, naked to the wind, go into their Sahara mode," Stevens writes.

A "Sahara mode" involves migrating sand dunes that could bury some highways and railroads, and wind-blown sand that would bombard productive fields downwind. Such migrating sand dunes have occurred in the past in the Great Plains—the Nebraska Sand Hills are the largest dune field in the Western Hemisphere—and they could occur again, even without readings-style Earth changes.

Reading 3651-1 seems to imply that Montana may one day have "much to do with many, many nations!" Thus, eastern Montana, as an outlier of safety-land Saskatchewan, together with a portion of Nevada, could be specified as safety lands in the same sense as "portions of . . . Ohio, Indiana and Illinois."

PART THREE
THE NEW MILLENNIUM

9

Humanity's Response to Earth Changes

In the spring of 1941, an apprehensive time for America as Europe was engulfed in World War II, Edgar Cayce gave a reading for friends who were worried about the period of years just ahead. We enter the reading near the middle:

> Strifes will arise throughout the period. Watch for them near Davis Strait in the attempts there for the keeping of the life line to a land open. Watch for them in Libya and in Egypt, in Ankara and in Syria, through the straits about those areas above Australia, in the Indian Ocean and the Persian Gulf.

Ye say that these are of the sea; yes—for there shall the breaking up be, until there are those in every land that shall say that this or that *shows the hand of divine interference, or that it is nature taking a hand, or that it is the natural consequence of good judgments.* [Emphasis added.]

But in all of these, let each declare whom ye will serve: a nation, a man, a state, or thy God?

For to Him ye must look for comfort that ye know that comes. All that is of a temporal nature, this—too must pass away; yet there remains the comfort for those who declare themselves, "Let others do as they may, but as for me, I will serve the living God."

                                          3976-26
                                     April 28, 1941

The italicized words offer three models of understanding Earth changes: divine intervention, nature taking a hand, and the natural consequence of good judgments. Let's consider each one.

DIVINE INTERVENTION

Some of the readings have implied divine intervention as an explanation of the forthcoming Earth changes. This is implicit in the following extracts:

What is needed most in the Earth today? That the sons of men be warned that the day of the Lord is near at hand, and that those that have been and are unfaithful must meet themselves in those things which come to pass in their experience.

                                           5148-2
                                      May 29, 1944

Q Regarding the general world conditions, is it likely that changes in the Earth's surface in the Mediterranean area will stop Italy's campaign against Ethiopia?

A Not at *this* particular period. This may *eventually* be a portion of the experience, but not just yet.

Q When is this likely to occur?

A As to times and places and seasons, as it has indeed been indicated in the greater relationships that have been established by the prophets and sages of old—and especially as given by Him, "As to the day and the hour, who knoweth? *No one,* save the Creative Forces."

Tendencies in the hearts and souls of men are such that these may be brought about. For, as indicated through these channels oft, it is not the world, the Earth, the environs about it nor the planetary influences, not the associations or activities, that *rule* man. *Rather* does man—by *his compliance* with divine law—bring *order* out of chaos; or, by his *disregard* of the associations and laws of divine influence, bring chaos and *destructive* forces into his experience.

For *He* hath given, "Though the heavens and the Earth pass away, my *word* shall *not* pass away!" This is oft considered as just a beautiful saying, or something to awe those who have been stirred by some experience. But applying them into the conditions that exist in the affairs of the world and the universe in the present, what *holds* them—what are the foundations of the Earth? The word of the Lord!

416-7
October 7, 1935

It seems probable that at the beginning of the obvious Earth changes, only a small percentage of the world's population may be expected to think in these terms, especially in the Western world. But even scientists, such as John Marr, M.D., who wrote (in the May, 1996, issue of *Caduceus,* with C. Malloy) an article entitled, "Epidemiologic Analysis of the 10 Plagues of Egypt," can sometimes see a relationship between biblical events and divine intervention. Dr. Marr says, "When God does something to humans, He translates it into things we can understand—with the exception of miracles." Or as Rabbi A. Weiss puts it, "God most often works within natural ways. But even if it looks natural, it has a supernatural imprimatur. That it happens when it happens, to whom it happens, indicates a higher power." (Both quotes are from A. Raver, *New York Times,* 4/4/96.)

NATURE TAKING A HAND

The Native American culture has long foretold of coming Earth changes that will result from Mother Earth being upset with the overexploitation and unbalancing of nature by Earth's greedy inhabitants and with the poisoning of air and water by man-made chemicals. Hopi medicine man Rolling Thunder spoke about this in his biography[1] by Doug Boyd:

> When you have pollution in one place it spreads all over. It spreads just as arthritis or cancer spreads in the body. The Earth is sick now because the Earth is being mistreated, and some of the problems that may occur, some of the natural disasters that might happen in the near future, are only the natural readjustments that have to take place to throw off sickness. A lot of things on this

land are like viruses or germs. Now, we may not recognize the fact when it happens, but a lot of the things that are going to happen in the future will really be the Earth's attempt to throw off some of these sicknesses. This is really going to be like fever or vomiting, what you might call physiological adjustment.

A prime example of pollution originating in one place and then spreading all over is found in nuclear power's greatest calamity, the explosion in the Ukraine of the Chernobyl power reactor in 1986. A minimum of 100 million curies of radioactivity was released that affected Belarus, Russia, Georgia, Poland, Sweden, Germany, Turkey, and even such distant lands as the U.S. and Japan. Now, scientists have determined that "floods are carrying radioactive waste, dumped around the reactor after it exploded . . . into water supplies hundreds of kilometers downstream . . . [putting] more than 30 million people at risk of being contaminated by radioactivity now sweeping downriver from the Chernobyl nuclear power station" (*New Scientist*, 3/23/96).

In a subsequent interview in the *Mother Earth News* (July/August 1984), Rolling Thunder offers a vision that has some similarities to those of Phylos:

. . . unlike some foreign gurus who claim they know the exact day and time *[of the Day of Purification]*—I can't say exactly when it will occur . . . I've had visions of the planetary cleansing, and it's truly a horrible thing. Unlike the natural disasters of the past—which involved either fire *or* water—this upheaval will involve both fire *and* water. Cities will be reduced to rubble, and most of the few survivors will turn on each other violently. The

whole thing will take some time though. There actually will be 40 years of purification ...

In a more European, yet metaphysical rendering, we have Rodney Collin's concept from his book, *The Theory of Celestial Influence,* that the different natural realms of Earth have their centers of gravity in different periods of the table of elements. "We may suppose, for example, that the true home of the unnatural radioactive elements of period 7 is in the inert core of Earth."[2] According to Collin, our use of radioactive elements and the creation of transuranic elements for destructive purposes represent a violation of Nature's design. Indeed, excess plutonium, the chief man-made explosive ingredient in nuclear weapons, has been piling up worldwide for over a half century, until now there are around 1,200 metric tons at scattered points over the face of the Earth. And to this we can add thousands more tons of legal and illegal uranium-235 and -238. In the cosmology of Rodney Collin, Nature herself may rebel at such an imbalance, because such radioactive materials belong to deeper realms of the Earth and because our creation of nuclear bombs courts natural disaster.

The worst of the Earth's environmental contamination, exclusive of Chernobyl, is from nuclear complexes and bomb test sites in the former Soviet Union and the United States. The worst of the contamination in the U.S. occurs at (1) Oak Ridge National Laboratory, Tennessee (Fig. 16), where one million curies of radioactive cesium and strontium were injected underground; (2) the Savannah River Plant, South Carolina (Fig. 16), where 900,000 curies of radioactive solutions of mixed fisson products were released to streams and surge ponds; (3) the Hanford Plant at Richland, Washington, where 700,000 curies of mixed-fission product solutions were released into soils and surface ponds; and (4) the Nevada Test Site in southern Nevada (northwest of Las Vegas), where repeated underground (and above-ground) nuclear bomb testing has left huge quantities of plutonium, cesium, and other radioactive material underground, and radioactive fallout has spread far beyond the test site. Interesting, isn't it, that two of these locations (2 and 4 above) of concentrated radioactive pollution that began to be generated about 1943-1945 coincide with parts of the U.S. ("the southern portions of Carolina, Georgia" and "the southern portions of Nevada," respectively) that were said clearly to be due for Earth changes in readings given *prior* to 1943? And the other two locations are in either the western portion of the U.S. (3) where the land will be "broken up" or, as with location 1, near the Eastern Tennessee Seismic Zone (Fig. 16) where the release of seismic energy is currently higher than anywhere else in the eastern U.S.

As we speculated would be the case for divine-intervention-type reasoning, probably only a small percentage of the Earth's population will think in terms of "nature taking a hand" when attempting to explain the first of the obvious Earth changes.

NATURAL CONSEQUENCE OF GOOD JUDGMENTS

How do modern societies like ours tend to regard natural disasters? Our materialistic, cause-and-effect-type thinking says: On a planet that is itself alive, safety is an illusion. Neither psychology nor religion offer much solace in dealing with larger-than-human natural events. Therefore, the best that we can do is to monitor Earth processes that can cause disasters, organize ourselves to mitigate their effects, and live with the results, knowing that we have done our best as a society to reduce losses.

To help mitigate calamities, several U.S. federal agencies in 1987 advocated an International Decade for Natural Hazard Reduction for the period 1990-2000.[3] The agenda for the program is instructive. It covers such topics as natural hazards knowledge, knowledge collection and dissemination, problem-focused research, and regional warning systems. Such a program, even if implemented in full, would be totally inadequate to deal with the scope of the Earth changes described in the Cayce readings.

The U.S. approach to dealing with natural disasters calls for all nations, through United Nations connections, to join in this worthy effort. This sort of a bureaucratic, intergovernmental effort is truly in the idiom of America's conglomerate or highly diversified corporate approach to "getting the job done." This exemplifies American national character such as Cayce addressed in an interesting pre-World War II reading.

> Each nation, each people have builded—by the very spirit of the peoples themselves—a purposeful position in the skein, the affairs not only of the Earth but of the universe!
>
> And the peoples of France, then, have builded a dependence and an independence that makes for the enjoying of the beautiful, a reverence for the sacredness of body, as well as the Creative Forces within the experiences of all.
>
> Just so is there the result in England. Just so is there the conglomerate force in America. Just so are there the domination forces in Japan, China. Just so in Russia is there the new birth, out of which will come a new understanding. Italy—selling itself for a mess of pottage. Germany—a smear upon

its forces for its dominance over its brother; a leech upon the universe for its *own* sustenance!

1554-3
March 25, 1938

What has this got to do with Earth changes being "the natural consequence of good judgments"? Only this, that the Cayce concept of "chief features" that characterize nations may enter into our formulation of moralistic reasons for Earth changes, as explained below. I believe that most people will explain the first of the forthcoming Earth changes in moralistic terms. This is because people want to read into disasters that cause human suffering a special meaning—that they are in the nature of a punishment inflicted by "good forces" for immoral behavior. I think that reading 3976-26 contrasts the divine-intervention explanation for Earth changes discussed earlier with an explanation which says that whatever is discordant with accepted ethical principles or the dictates of conscience (immoral conduct) may, as a natural consequence, fall victim to a "good judgment"; that is, a judgment by the cosmic forces of good.

MORALISTIC REASONS FOR EARTH CHANGES

Now let's look at hypothetical examples of why, metaphysically speaking, Earth changes could occur in the near future at specific places in two countries, the U.S. and Japan. To facilitate our search, we'll look for morally despicable human activities or tendencies in places where the readings have already identified clear potentials for significant Earth changes. A dangerous morality-game, you say? You've no right to judge others, you say? I agree that we should not judge individuals, but here we're dealing with moral judgments pronounced on *group* behavior. And I'm quite aware that:

> **We live in times when it is rude**
> **To have a moral certitude.**
>   **Richard F. Barrett**

As for the United States, start with the comment in reading 1554-3 that there is "the conglomerate force in America." There can, of course, be good conglomerates, like the American Red Cross, and there can be bad conglomerates, such as elements of the U.S. government-military-industrial complex. An example of the latter is disclosed in a report by the Nuclear Weapons Cost Study Project Committee, published in the November 1995 issue of the *Bulletin of the Atomic Scientists:*

> **Since the government first began work on the atomic bomb in 1940, the U.S. nuclear arsenal has cost about $4 trillion in 1995 dollars—or approximately three times more, in 1995 dollars, than was spent on procurement for all of World War II ... The constitutional systems of checks and balances was rarely applied to the nuclear weapons program. A cloistered bureaucracy, largely unaccountable to Congress, managed the program, and Congress frequently neglected its oversight function, allowing decades of wasteful spending. Production took precedence over health and safety, leading to environmental degradation and the endangerment of nuclear weapons personnel and the civilian population. Pervasive and often unnecessary secrecy drove up costs, prevented the American people (and sometimes Congress) from knowing the full extent of nuclear weapons programs, and helped undermine the people's trust in government.[4]**

But even worse, a talent pool of excellent scientists and engineers was diverted from constructive work to the development of enormously destructive weapons, far more than would be needed for national defense, even if such weapons could be construed as defensive. This program has burdened America with 70,000 nuclear warheads, 67,000 missiles, 4,000 bombers, and 59 strategic missile submarines. To achieve this colossal power to destroy required 1,041 nuclear test explosions.

The bulk of weapons design and testing was conducted by the Department of Energy's three weapons laboratories: Sandia National Laboratories in Albuquerque, New Mexico; Los Alamos National Laboratory near Santa Fe, New Mexico; and the Lawrence Livermore National Laboratory in Livermore, California. The nuclear weapons production complex, which once included several hundred facilities, today includes 14 production reactors (in Washington State and South Carolina), eight separation and reprocessing plants, and 239 underground storage tanks for high-level radioactive waste (in Washington, Idaho, and South Carolina). Tons of machined plutonium for nuclear weapons are in storage at Rocky Flats, 16 miles from downtown Denver, Colorado. We have already covered the potential for Earth changes in the western states where the readings said the Earth would be "broken up." The bulk of America's nuclear-program facilities reside there. The Savannah River Plant is located in (or near) the part of South Carolina due for "disappearance" (by subsidence), according to reading 1152-11. Oak Ridge National Laboratory, founded in 1943 to construct an atom bomb, is located close to the increasingly active Eastern Tennessee Seismic Zone. (Fig. 16)

The most egregious example of financial waste in any of the weapons laboratories can be found at Lawrence

Livermore National Laboratory. In addition to its share of the enormous amount of funds spent wastefully on nuclear weapons development and testing, Livermore squandered the bulk of the funds designated for the missile shield program, the Strategic Defense Initiative (S.D.I., or "Star Wars"). "Billions went into implausible research projects at a single well-connected office at . . . [Livermore]," according to commentator A.M. Codevilla. "One [such project], the 'X-ray laser,' promised to direct the energy of hydrogen bombs thousands of miles with great precision. Another promised to generate huge 'free electron' laser beams and transmit them through the atmosphere without significant loss. Livermore also promised 'smart rocks' and 'brilliant pebbles,' little rockets that would put themselves in the way of incoming missiles."[5] This certainly sounds like reincarnated Atlanteans from Poseidia at work here in America.

Now the bad news is that Star Wars is back. Livermore's emeritus scientist, Edward Teller, the legendary father of the hydrogen bomb, who sold then-President Reagan on the Strategic Defense Initiative concept in the early '80s, admitted in 1995 that the original S.D.I. "could never have defended 'against an all-out attack by the Soviet Union.' But Star Wars makes sense now, he argued, because there are fewer missiles in the world."[6] Yet after $30 billion already spent on S.D.I., mainly at Livermore, there are still no weapons.

Lawrence Livermore National Laboratory sits just east of the Calaveras and Hayward faults in California. The Hayward fault is considered to be "the most likely source of one or more major earthquakes in the San Francisco Bay Area in the next few decades."[7] But if the Earth-change readings are correct, San Francisco [and probably contiguous Bay-Area cities] will be destroyed soon, near the beginning of the period of global Earth changes.

Could the destruction of Livermore and environs be considered to be a "natural consequence of good judgments"? One might reason so, knowing what is now known about the waste of national treasure there and the out-of-all-proportion diversion of scientific and engineering talent to destructive weapons development at Livermore. One's reasoning here would depend on one's governing understanding or philosophy.

Our above conjectures raise many questions. One concerns the role that individual responsibility plays in the following moral assertion: *Here at the end of the age, any large group that has conceived of or is implementing morally reprehensible actions can be subjected to an Earth-change type of judgment.* But isn't it acceptable to join together with others to help the nation defend itself? The readings would say "yes," if the projects are truly defense oriented. The nagging question then becomes, Is it all right to join others involved in a grossly selfish program to develop weapons of mass destruction that is all out of proportion to a society's defense needs?

As for the role of the individual, we *know* it is important with whom or with what group one associates. And if the ultimate agenda of the group is disharmonious with the higher purposes of humanity, will the individual involved with such a group be protected if he merely goes to church on Sunday and helps coach Little League on Wednesday? It's not that we have the answers here. They are for the mind of God. Rather, we are raising questions that move our thinking into new and unusual avenues that may be of value later on, in the new age unfolding before us.

MORALISTIC REASONS FOR A
HYPOTHETICAL EARTH CHANGE IN JAPAN

The national chief feature that we are dealing with in this section may be the same as the one given in 1938 (reading 1554-3): "*Just so are there the domination forces in Japan, China.*" We now set the stage for our conjectures about a monumental Earth change in Japan by referring to some background facts provided by the Nuclear Control Institute (NCI), nuclear proliferation's worldwide watchdog (as posted on November 21, 1995, on website http://www.nci.org/nci/):

> **There will soon be more atom-bomb material in civilian nuclear energy programs than now exists in nuclear arsenals. Plutonium is being extracted from reactor wastes for use as fuel in nuclear power plants. Uranium has been enriched to weapons grade for use as fuel in research reactors. These civilian, bomb-grade fuels are unnecessary because nuclear power and research reactors can be run without them. Theft of less than 20 pounds of plutonium or 40 pounds of bomb-grade uranium from civilian programs could result in a 35-kiloton nuclear explosion. Nations that stockpile tons of plutonium and bomb-grade uranium for peaceful purposes could convert these fuels into nuclear weapons at any time.**

In 1994 inspectors from the International Atomic Energy Agency (IAEA) were "politely asking the Japanese to physically produce about 70 kilograms (155 pounds) of plutonium that the operator of a plutonium fuel plant says are stuck in the processing equipment. What worries the inspectors is that the amount reported to be 'held up' in the process stage of the plant, enough for 10 to 20 bombs, is running several times larger than antici-pated in the plant's design . . . [and] the margin of error is so large that it can mask deliberate diversions of plu-tonium. Imagine what lies ahead when commercial plants 30 times the size of the Japanese facility are built."[8]

The author of this quote, Paul Leventhal, president of NCI, goes on to say that he once asked an IAEA official if the Japanese could beat the agency's safeguards system in a plutonium plant. He was told only that "No system is perfect." Leventhal continues, "Japan's neighbors are not amused. South Korea has put the U.S. on notice that it will demand the same permission to recover pluto-nium from U.S.-supplied nuclear fuel as we gave Japan if the Japanese go ahead with their commercial pluto-nium program. North Korea accuses Japan of pursuing nuclear weapons and uses Japan's program to justify its own. The prospects of a 'plutonium race' in this region of ancient enemies looms large."[8]

Japan has moved beyond the pilot-plant stage and is now building its first commercial plant for reprocessing highly radioactive, spent uranium fuel from its nuclear power reactors. The commercial plant is being constructed in Rokkasho,[9] which is located on the extreme northeast-ern tip of Honshu. (See the letter R on Fig. 17.) The new plant is designed to supply about 4.8 tons of plutonium annually. Removing the plutonium from spent fuel may have dangerous consequences. "The risks involved make no economic sense because the recycled pluto-nium is four to eight times more expensive than ura-nium fuel. Moreover, uranium is in abundant supply, and nuclear reactors are perfectly capable of running on low-enriched fuel that cannot be used in weapons."[8]

The Rokkasho recycling plant appears to be right in the middle of the part of Japan that is destined to "go into the sea" (3976-15). Only the timing of the event seems to be at issue.

As history shows, Japan in the 1930s was controlled by what Cayce calls the "domination forces," who wished to dominate all Asia. Domination forces led to the Rape of Nanking, the Bataan Death March, the invasions and occupations of Japan's neighbors to the west, and to the attack on Pearl Harbor. Could this contribute to Japan's going into the sea decades later? Are domination forces still active in Japan today? Or have the Japanese finally overcome the chief feature of their society that led them to try to dominate their neighbors militarily? Perhaps the terrible punishment wreaked on the Japanese people during World War II expiated the sins of the souls that then constituted Japanese society. But what about tendencies in that society today?

What about the Aum Shinrikyo religious sect that introduced nerve gas into Tokyo's subways? The sect had planned even larger terroristic activities in Japan and beyond. A Buddhist nun said that Aum symbolizes an illness that has been growing in Japan ever since World War II, in which society there appears disenchanted, without ideals or even cynicism. Japanese author Reiko Hatsumi wrote[10] that "Japan lacks something else most other cultures have: established religions that have stood the test of time and offer a moral standard. Even if one deviates from this standard, one knows it is there. Shinto has no code of ethics or dogma. Buddhism here [in Japan] has degenerated into a set of rites with which we are buried. Christianity flowered briefly in the Middle Ages, but was ruthlessly suppressed."

In November 1995, the head of the Japanese Finance Ministry's international finance bureau is reported[11] to have said that Japan might forge an alliance with China against the U.S. if Washington maintained a "coercive attitude" toward Japan. (Recall that China was the other nation mentioned in 1938 [reading 1554-3 above] as a possessor of "domination forces." And China continues to try to dominate its neighbors Taiwan and Tibet.[12])

As for Japan's possible intention to stockpile atom-bomb materials, the IAEA "cannot even get Japan to clean out a nuclear-fuel plant promptly and come up with 150 pounds of plutonium (enough to make a dozen Nagasaki bombs) that Japan insists is stuck in the pipes."[13] If Japan tries to go nuclear and succeeds in making a bomb today where she failed in the past,[14] will not moralistic people everywhere begin to think that a "good judgment" might well be needed for that nation?

There is more than enough information on real-world associations between evil deeds and intentions in specific nations and the propensity for Earth-change-type judgments there. Other examples abound, among them Germany, which in 1995 announced plans to fuel a research reactor at Garching, near Munich, with bomb-grade plutonium. In 1996 the Germans "rejected American demands that the reactor be remodeled to use low-enriched uranium."[15] America's demands were perhaps undermined by its own actions: in 1995, the U.S. Department of Energy "reprocessed some fuel in a chemical factory at the Savannah River Site," in the part of South Carolina that reading 1152-11 said would "disappear." This according to Matthew Wald in the *New York Times*. He writes, "The United States, which had given up recovering plutonium from spent nuclear reactor fuel and had urged Russia, Japan, and others to do the same to help curb the spread of nuclear weapons, is about to resume the practice on a small scale."[16]

A hydrogen-bomb explosion simulates solar energy itself, and whether such divine energy can be handled safely by humanity is perhaps the chief riddle of our age.

Phylos said that "America, the Glorious, together with the rest of the world" will meet worse woe than Atlantis did, "though not by water but by fire." If so, and if Phylos is correct that "man is his own judge and executioner," then the *fire* that Phylos is talking about could be the fire of nuclear explosions. "Los Angeles, San Francisco, most all of these will be . . . destroyed before New York even," said Cayce. Who needs divinely managed, nature-demanded, or moralistically determined Earth changes to discipline a God-indifferent world when nuclear weapons are so convenient and capable for the task?

Although the Cold War is over, international hostilities remain. America's nuclear-tipped missiles can be reprogrammed within minutes to target Chinese and other "enemy" populations and real estate, and vice versa for their missiles. Despite efforts to convince the world's nuclear powers to sign a comprehensive test ban treaty, China and France tested nuclear warheads in 1995. If India follows suit, its neighbor, Pakistan, with which it has tense relations, may also test a nuclear weapon. Neither country has signed the Nuclear Nonproliferation Treaty. According to Gary Milhollin, director of the Wisconsin Project on Nuclear Arms Control in Washington and a leading civilian expert on the spread of nuclear weapons, Indian bomb testing will have a snowball effect and jeopardize the possibility of worldwide agreement on a comprehensive test ban treaty.[17] If there is an international rivalry today that is so hostile and unstable as to result in the launching of nuclear weapons out of anger, it is the rivalry between India and Pakistan. These countries are home to about one-fifth of the world's population, and even a "limited" exchange of nuclear warheads between them would present an humanitarian and ecological disaster of near-biblical proportions.

Perhaps, then, Earth changes sufficient to stop the

nuclear madness are preferable to eventual nuclear warfare. Perhaps both warfare and Earth changes will take place and Rolling Thunder will be right when he says that this time the world will be destroyed by both fire (from volcanoes and nuclear explosions?) *and* water (the readings' prediction of submergence of most of Japan, southern Carolina and Georgia, and the coastal WWII battlefields?).

In the long journey we're taking to discover truth, we may need to experience Earth changes at some point. It is hoped that if humanity fails to make headway, the Universal Forces will prevent power-hungry leaders from perpetuating nuclear war, perhaps by initiating Earth changes. How we respond to such changes will be all important. The following readings about our attitudes in the face of adversity and about how to head off the consequences of dangerous intergroup and international tensions are well worth considering here.

Q Any advice regarding . . . the attitude for us to hold?

A The attitude of all should be:

"Thy will be done, O God! And let me find myself content with that I cannot change, and to change that which I may that will be in closer keeping with Thee."

1100-37
June 23, 1943

Q What can we do to counteract such serious happenings?

A Make known the trouble—*where it lies; that they who have forgotten God must right about face!*

3976-26
April 28, 1941

As we near the end of this particular age, are very self-ish, wicked, or generally God-indifferent types of people becoming attracted to locations of impending disaster? Aphorisms like "You attract your life" or "You'll get the Earth change you deserve" come to mind. Can we, by right living, really hold together the ground under our feet? Or is this just an allegorical construct to help us understand the importance of aspects of our internal life? Or are both possibilities correct?

10

A New Age, A New Understanding

CHANGES TO COME IN THE AQUARIAN AGE

It would be momentous if all of the Earth changes predicted in the Cayce readings were to occur. It would be even more momentous if some or all of the changes destructive to society could be mitigated by human understanding, followed by work toward spiritual attunement. Could this be one of the promises of the Aquarian age, in which a God-seeking humanity will have a capability of understanding what really is happening? That's what one reading suggested.

As has been indicated, we will begin to understand fully in '98.

1602-3
September 22, 1939

And individuals may realize their potential for influencing positively their material environs and the forces of nature herself. Consider the following two readings that speak to these possibilities:

Q What will the Aquarian age mean to mankind as regards physical, mental and spiritual development? . . .

A Think ye this might be answered in a word? These are as growths. What meant that awareness as just indicated? In the Piscean age, in the center of same, we had the entrance of Emmanuel or God among men, see? What did that mean? The same will be meant by the full consciousness of the ability to communicate with or to be aware of the relationships to the Creative Forces and the uses of same in material environs. This awareness during the era or age in the age of Atlantis and Lemuria or Mu brought what? Destruction to man, and his beginning of the needs of the journey up through that of selfishness.

Then, as to what will these be—*only* those who accept same will even become aware of what's going on about them! How few realize the vibratory forces as create influences from even one individual to another, when they are even in the same vibratory forces or influence! And yet ye ask what will the Aquarian age bring in mind, in body, in experience?

Q Is the Aquarian age described as the "Age of the Lily" and why?

A The purity. Only the purity as it represents will be able to comprehend or understand that awareness that is before those who seek the way.

1602-3
September 22, 1939

Two months later Cayce gave a reading for the same person, who asked for clarification of the many changes to come. Hugh Lynn Cayce conducted the reading.

1 HLC: You will have before you the entity, [1602], also her mental and material affairs. It has been indicated, through various sources, that the period immediately ahead [post-1939], is one of change in many fields—politically, economically, and geologically. Will you clarify these predictions for me, and give directions for me at this time? answering questions which have been prepared?

2 Edgar Cayce: Yes, we have the entity here.

3 As in relationship to changes—these are indicated not only through prophecies but through astrological aspects, as well as the thought and intent of persons and groups in high places; bringing about these things, these conditions, in what might be said to be the fullness of time.

4 However, since the advent of the Son of Man in the Earth, giving man an advocate with the Father, there has been an influence that may counteract much of that which has been indicated that would come as retribution, or in filling the law of an evolution of ideas and the relationship of material things to the thoughts and intents of individuals and groups.

5 Then, as to whether the hearts and minds of individuals or souls (who were given authority

concerning the laws of the universe) are fired with the thoughts of dire consequences or those things that bespeak of the greater development of a spiritual awakening, is still in the keeping and in the activities of individuals who—as this entity—have caught a glimpse, or an awareness, of that which is in the making, in the affairs of state, nation, and nations, and the universe, as related to the conditions upon the face of Mother Earth.

6 There enters much, then, that might become questions as respecting that which has been foretold, or prophesied, as well as respecting the activities of groups and individuals who have acted and who are to act as a counterbalance to these happenings in the Earth.

7 In the first premise—know what was the cause of indifference, or sin, entering material manifestations. Was it the purpose by God that such should be, or by the Godhead? or was it that this force or power seeking expression found— with the expression—that there came the forces of positive and negative?

8 And with same the awareness of one influence or force, taking certain courses or directions, became negative.

9 The others became the greater positive.

10 Thus in the experience of souls through their evolution in the material things of the Earth, there has been brought just that same effect in the material affairs of the souls active in expressing or manifesting at this particular period or sphere of development.

11 Much of just this comprehending is indicated in some of those records that are now becoming more and more a part of man's experience,

or awareness; in that the cosmic or universal or spiritual laws are bringing same into that category or phase of experience where they become a part of individual experience.

12 This may be indicated from the records in the rocks; it may be indicated in the pyramids,—man's attempts to leave a sign to those who, in the spiritual comprehension of material associations in spirit, would interpret that which had been, that which is, and that which was to be.

13 Hence it is seen that these are interpretations that become a matter of the consciousness of the individual so making same.

14 Or, to return to the first premise, it depends upon which line is taken by such an individual making such interpretation; whether a pessimistic or an optimistic, or a positive or a negative; or (by negative we mean) one that sees the world, as related to the Earth and its position in the universe, being damned irrespective of what souls do about same—taking little or no account of the words, the promises, yea the activities of Him. He manifested in the Earth that as would bring to the seeker an awareness of the constructive influence of same.

15 This may be indicated or seen in the record according to the Book—which is as a sign, a guide to those who seek to know His ways, His purposes to man.

16 These interpretations of the promises, the pledges taken and given in the lives and activities—or during the phases of a sojourn of an individual soul, must be taken into account.

17 And then these indicate as to what is to come to pass, even through these periods of the Earth's

journey through space; "catching up," as it were, with Time.

18 And then the soul realizes—in his search for his Maker—the patience that was, is and will be manifested in Him; He that is the way, the truth and the light.

19 *Again* the interpretation of the signs and the omens becomes an individual experience. And each soul—as this entity—then is given the privilege, the opportunity to *live* such an activity in its relationships to its fellow man; filling, fulfilling, and interpreting that which has been indicated, in such measures and such manners as to bring hope and not fear, peace and not hate, that which is *constructive* and *not* destructive, into the lives and minds and hearts of others.

1602-5
November 28, 1939

Reading 1602-3 reminds us of the importance of being aware of subtleties, such as the vibratory forces that create influences from one individual to another. And awareness of the influences of the Aquarian age will require an inner effort unlike any we are used to today. The above two readings are almost a call for developing one's intuition or a new organ of perception,[1] metaphorically speaking. Such a faculty can function, however, only in one capable of exercising the purity of thought and emotion that comes from actively and correctly seeking the way. And this requires right orientation, proper efforts, and time.

The relationship to conditions on Earth of the thoughts and emotions of certain souls "who were given authority concerning the laws of the universe" are mentioned in paragraph 5 above. The reading seems to imply that the way in which certain souls think about or judge ongoing trends in human behavior may result in either dire consequences or "things [activities?] that bespeak of the greater development of a spiritual awakening." Paragraph 6 suggests a dependence of activities on Earth (including presumably Earth changes) upon specific groups and individuals who have acted in the past and who are to act in the future, "as a counterbalance to these happenings [untoward events mentioned in the question of paragraph 1] in the Earth." Whether these groups and individuals are all incarnate at any given time is anybody's guess, but it sounds as though some are "individuals," meaning alive now, and some are "souls," or disincarnate now. Phylos (see Chapter Four) comes to mind in connection with the latter type.

Paragraphs 11-13 make the interesting point that the interpretations that we make of universal or spiritual laws coming upon the Earth, as well as those records in the rocks and pyramids, are a matter of the consciousness of the individual doing the interpreting. And the individual's consciousness here is a strong function of prior experience "in the material things of the Earth" (paragraph 10). Thus, some will be better able than others to interpret the universal or spiritual laws of the new age beginning to be felt in 1998. Paragraph 14 amplifies somewhat the nature of the pessimistic or optimistic premises that may be taken by these special souls. But as we read onward to paragraph 19, we realize that each of us may be a special soul, *if* we seek "to bring hope and not fear, peace and not hate, that which is *constructive* and *not* destructive, into the lives and minds and hearts of others."

HIGH SOULS RETURN

Other readings further describe the promise of the new age that begins in 1998:

Then began the laying out of the pyramid and the building of it, through use of those forces that were able to bring the materials from those very mountains that had been a place of refuge for Ra Ta and where certain memorials had begun. The various sections were laid out not only for the receiving of that which had been offered in the Temple Beautiful, on the various altars for the activities of an individual's innate self, but for the place of initiation of the initiates who were to act in the capacity of leaders in various activities. This building on the pyramid lasted for a period of one hundred years, as now termed. It was formed according to that which had been worked out by Ra Ta in the mount, as related to the position of the stars about which this particular solar system circles in its activity—going toward what? That same name to which the priest was banished, Libya—or the constellation of Libra. Is it not fitting, then, that these must return? This priest Ra, now Edgar Cayce, may develop himself to be in that position or capacity of a liberator of the world, in its relationships to individuals in those periods to come; for he must enter again at that period— or in 1998.

Paraphrased by Hugh Lynn Cayce
from reading 294-151, July 29, 1932

In this same pyramid did the Great Initiate, the Master, take those last of the Brotherhood degrees with John, the forerunner of Him, at that place. As is indicated in that period where entrance is shown to be in that land that was set apart, as that promised to that peculiar peoples, as were rejected—as is shown in that portion when there is the turning back from the raising up of Xerxes as the deliverer from an unknown tongue or land, and again is there seen that this occurs in the entrance of the Messiah in this period—1998.

5748-5
June 30, 1932

We learn in these two readings that not only will Edgar Cayce be born again into the Earth in 1998, but that—in some way—the Messiah will also enter then. And what will be the nature of the Messiah's entry in 1998?

Q Please explain what is meant by "He will walk and talk with men of every clime." Does this mean he will appear to many at once or appear to various peoples during a long period?

A As given, for a thousand years He will walk and talk with men of every clime. Then in groups, in masses, and then they shall reign of the first resurrection for a thousand years; *for this will be when the changes materially come.* [Emphasis added.]

In the manner as He sat at the peace conference in Geneva, in the heart and soul of a man not reckoned by many as an even unusually Godly man; yet raised for a purpose, and he chose rather to be a channel of His thought for the world. So, as there has been, so will it be until the time as set. As was given of Him, not given to man to know the time or the period of the end, nor to man—save by their constituting themselves a channel through which

He may speak. To be sure, man and woman alike; for, as given from the beginning, they are one. Not as man counts oneness from the material viewpoint. Rather from that as He gave, "Where are those thy persecutors?" "No man, Lord." "Neither do I condemn thee. Sin no more." In the searching out of those who, where, and when He speaks, or has spoken among or with, or through men, "By their fruits ye shall know them." They that bring more righteousness are the children of faith, hope, charity. These three; and they do so in the *material* world, the Father, the Son, the Holy Spirit. Be thou, then, a channel that may oft walk with Him that gave not of else than, "Let not your heart be troubled; neither let it be afraid. Let not thine right hand know what thy left hand doeth." Rather giving self to seeking, day by day, to *know* the will of the Father as was manifest in Him, and may be manifest in Thee, for He will not leave thee desolate, but will come to thee—but not unless invited . . .

364-8
April 15, 1932

HOW CAN WE PREPARE FOR THE NEW AGE?

Approaching the changes introduced by the Aquarian age we must continue to ask, "What can I do about myself?"

In his book, *The Commanding Self,*[2] that addresses this and related subjects, the modern Sufi exponent, Idries Shah, defines the commanding self as "that mixture of primitive and conditioned responses, common to everyone, that inhibits and distorts human progress and understanding." Shah cites as examples of this "self" greed, anger, and pride, characteristics that blind us from awareness and from fulfilling our potential.

Finding the correct study method is one of the keys to success in any field, no less in our spiritual development. And truly effective learning, whether individually or in groups, requires expert guidance, specific to the formulation of one's chosen developmental way. Shah cautions against following three erroneous paths commonly taken today—the scholastic-study, the guruistic, and the morbidly religionist paths. He says that these three erroneous paths are effectively cults, in which legitimate spiritual teachings have fallen into the hands of personality types who distort the spiritual teachings' outer shapes.

People can learn in a group, but only when the group functions as a true learning group and does not degenerate—as an originally well-intentioned study group often does—into a mere social organization. To prevent this happening, some groups have made concerted efforts to study the social psychology of human groupings, while at the same time trying to carry out the objectives of the study group itself (say, for example, self-realization for each member). Reviewing information in books that deal with the dynamics of groups of randomly collected people, like the findings in W.J.H. Sprott's *Human Groups*,[3] has been found helpful. Also worth recommending to anyone motivated to evaluate spiritual groups is a monograph, *Evaluating Spiritual and Utopian Groups*,[4] by Arthur Deikman, a clinical professor of psychiatry at the University of California, San Francisco, who has made a particular study of the relationship between the mystical tradition and modern psychotherapy.

The Association for Research and Enlightenment (A.R.E.) has developed a worldwide network of study groups to help people begin to realize their spiritual potential. These are the so-called Search for God groups. When in 1931 a Norfolk group that had been attending

Edgar Cayce's Sunday afternoon lectures asked him to start a group that would study his work, he agreed and over the next 11 years he met with the group and gave 131 readings. The guidance in these readings became the basis for a study manual consisting of 24 lessons. The basic information for each lesson has been published as *A Search for God, Books I and II.*[5] A companion text is *Step-by-Step,*[5] a guidebook to the study and practice in *A Search for God, Book I.*

As for aligning oneself with the Association for Research and Enlightenment—or with any other association for that matter—the choice is always ours, as free agents. Consider the answers given to a woman who asked Cayce:

Q Are there special groups with which I should become affiliated, in doing the work that God is urging His people to do now, impersonally?

A May be a member of an organization, but rather than as a member of a group be a helpmeet to *all* groups!

Q [Should I be a member of the] Association for Research and Enlightenment, Inc.?

A This is identified with the Christian forces. Hold fast to the basic principles in that phase of same. There may be help there, but not as [one] identified with same. This is the identification— the Christlike principle!

1152-11
August 13, 1941

Now while this was advice for a specific individual, the matter of one's orientation toward any group is important, and the approach given in this reading is undoubtedly true for anyone.

In conclusion, here are a few more reading extracts for us to ponder and then, perhaps, to act upon positively.

> [How] to meet same [social turmoil, strife between capital and labor, and a division in America]? Only that each soul turns not to self alone and cry for strength, but that each soul *lives* in such a manner that there may be the awakening to the needs, the purposes, the causes for the nation to come into existence! That such [strife, etc.] is, and to be, a part of the experience of America is because of unbelief!
>
> 3976-24
> June 16, 1939

> It is also understood by some that a new order of conditions is to arise; that there must be many a purging in high places as well as low . . . there will then come about those circumstances in the political, the economic and the *whole relationships* where there will be a leveling—or a greater comprehension of this need [emphasis added].
>
> 3976-18
> June 20, 1938

> Q How does tribulation work patience . . . ?
> A As has been indicated in that "Whom the Lord loveth He chasteneth," and purgeth every one; for corruption may *not* inherit eternal life, and must be burned up. Know that thy God is a consuming fire, and must purge every one, that ye may enter in. In patience does one overcome.
>
> 262-26
> August 21, 1932

. . . we are our brother's keeper . . . if those in a position to give of their means, their wealth, their education, their position, *do not* take these things into consideration, there must be that leveling that will come [emphasis added].

3976-19
June 24, 1938

As the Spirit of God once moved to bring peace and harmony out of chaos, so *must* the Spirit move over the Earth and magnify itself in the hearts, minds and *souls* of men to bring peace, harmony and understanding, that they may dwell together in a way that will bring that peace, that harmony, that can only come with all having the *one Ideal;* not the one *idea,* but "Thou shalt love the Lord thy God with all thine heart, thy neighbor as thyself!" This [is] the whole law, this [is] the whole answer to the world, to each and every soul. That is the answer to the world conditions as they exist today [emphasis added].

3976-8
January 15, 1932

The future of the world—and certainly that of the human race—may depend on our response.

✳ ✳ ✳

ADDENDUM

UPDATE FOR 1997

It has been over a year since the first edition of this book was published. My earlier interpretations of the readings, and their correlations with geophysical findings and societal trends, need to be updated. Have new data and interpretations changed the book's conclusions in any way? No. Now, as much as ever, it seems possible that the Earth may soon be physically changed in unprecedented ways—and that humankind will grow enormously in the understanding of its origins, its destiny, and the methods by which people of humility may develop, psychologically and spiritually, in the New Age.

Updates that follow are sequenced according to the section or chapter order found in the book. Page references given in the updates are inserted to help the reader find related sections in the main text.

CAYCE'S PSYCHIC VISIONS
Revisiting the 1934 Halaliel Reading

As mentioned on p. 19, reading 3976-15 accounts for eight of the twenty-four Earth-change predictions covered in this book. They are among the most catastrophic of the predictions as well and were dictated by "archangel Halaliel," said to be a leader of the heavenly host and one who is "Not the Christ, but His messenger, with the Christ from the beginning, and is to other worlds what the Christ is to this Earth" (262-71). At the same time, there seems to be a significant error in Halaliel's answer to one of the questions asked in the reading. The seeming inconsistency in Halaliel's response, and the fact that most members of Cayce's Norfolk Study Group No. 1 declined an offer of Halaliel's help as teacher and guide in the group's deliberations,[1] has caused other interpreters of the readings to dismiss as suspect all of Halaliel's Earth-change predictions.[2]

As I wrote on p. 18, "But what about the accuracy of the Halaliel reading? Cayce was asked what changes could be expected in that year, 1934. The reply was startling . . . " I then went on to quote six predictions that clearly did not happen in 1934. Then, I speculated that Halaliel was either wrong or had simply ignored the "this year" specification in the question that resulted in the predictions. I suggested that the "this year" qualifier translated to "this period," 1958 to 1998, which had been mentioned earlier in Halaliel's discourse.

Further study of the Halaliel reading allows for another, more convincing explanation of the seeming inconsistency. Look first at the exact wording of the question. It is: *"What are the world changes to come this year physically?"*

We must ask, does the word "this" mean "the present," as in "What changes are to come in the present year?" Or does "this" refer to some word or phrase given earlier in the reading?

The antecedent of "this," in "this year," is most likely "the acceptable year of the Lord" being discussed two paragraphs earlier in 3976-15. The following unbroken quote from the reading shows the connection, and the two times that the word "year" is used are italicized purposely.

10 Who shall proclaim the acceptable *year* of the Lord in him that has been born in the Earth in America? Those from that land where there has been the regeneration, not only of the body but the mind and the spirit of men, THEY shall come and declare that John Peniel[3] is giving to the world the new ORDER of things. Not that these that have been proclaimed have been refused, but that they are made PLAIN in the minds of men, that they

may know the truth and the truth, the life, the light, will make them free.

11 I have declared this, that has been delivered unto me to give unto you, ye that sit here and that hear and that see a light breaking in the east, and have heard, have seen thine weaknesses and thine faultfindings, and know that He will make thy paths straight if ye will but live that YE KNOW this day—then may the next step, the next word, be declared unto thee. For ye in your weakness [pause] have known the way, through that as ye have made manifest of the SPIRIT of truth and light that has been proclaimed into this Earth, that has been committed unto the keeping of Him that made of Himself no estate but who brought into being all that ye see manifest in the Earth, and has declared this message unto thee: "Love the Lord thy God with all thine heart," and the second is like unto it, "Love thy neighbor as thyself." Who is thine neighbor? Him that ye may aid in whatsoever way that he, thy neighbor, thy brother, has been troubled. Help him to stand on his own feet. For such may only know the acceptable way. The weakling, the unsteady, must enter into the crucible and become as naught, even as He, that they may know the way. I, Halaliel, have spoken.

12 Q What are the world changes to come this *year* physically?

A The Earth will be broken up in many places. The early portion will see a change in the physical aspect of the west coast of America. There will be open waters appear in the northern portions of Greenland. There will be new lands seen off the

Caribbean Sea, and DRY land will appear. There
will be the falling away in India of much of the
material suffering that has been brought on a
troubled people. There will be the reduction of one
risen to power in central Europe to naught. The
young king son will soon reign [see par. 13-A be-
low]. In America in the political forces we see a re-
stabilization of the powers of the peoples in their
own hands, a breaking up of the rings, the cliques
in many places. South America shall be shaken
from the uppermost portion to the end, and in the
Antarctic off of Tierra del Fuego LAND, and a strait
with rushing waters.

13 Q To what country is the reference made regarding the young king?
  A In Germany.

It's always possible to assume that question 12 had
been formulated prior to the start of the reading and that
the question really *was* referring to 1934, because the
reading was given in January, at the beginning of that
year. But we can just as well assume that an archangel
may not like to have his train of thought interrupted. And
when the question was asked, Halaliel had just been
speaking about the acceptable year of the Lord.

Buttressing my argument that Halaliel's answer (12-A
above) was not referring to a typical 12-month year, consider that Hitler's rise and fall, predicted in answer 12-A,
required 11 years *after 1934* to be realized. Furthermore,
it has taken more like 50 years after 1934 for there even
to begin to be "the falling away in India of much of the
material suffering" there, also mentioned in 12-A. By extension of these observations, we have every right to assume that the Earth changes mentioned above in 12-A,

and in the rest of Halaliel's reading (see p. 18), are yet to come.

The "acceptable year of the Lord" in the Halaliel reading seems to coincide with the period of time mentioned in reading 364-8 (p. 235) that says that the changes will materially come over the 1,000-year period of the first resurrection [when] "He will walk and talk with men of every clime." And, as mentioned on page 23, the Messiah is to enter "in this period—1998" (5748-5, p. 235). Thus, there seems to be a convergence of readings that suggests that the catastrophic Earth changes mentioned in 3976-15 are to occur in 1998 and beyond.

Changes to Occur "In the Next Generation"

Two readings warned that destructive forces would come both to New York City and to "the land" sometime in the next generation. The first reading, given in 1941 (p. 99), is:

Q I have for many months felt that I should move away from New York City.

A This is well, as indicated. There is too much unrest; there will continue to be the character of vibrations that to the body will be disturbing, and eventually those destructive forces there—though these will be in the next generation.

The second reading, 3209-2, was given on December 31, 1943.

Before that the entity was in Atlantis when there were the periods of the first upheavals and the destructions that came to the land—as must in the

next generation come to other lands.

Various other interpreters[4,5] of the above two Earth-change readings have expressed a concern that the changes predicted to occur "in the next generation" will now never come to pass because the predictions have run out of the time allotted for their realization. The earliest of the two readings was given in 1941, and if the Earth changes are to start *in* the next generation, let's do the arithmetic to evaluate the latest date by which the changes can occur for the 1941 reading to still be in force.

Start with the dictionary definition that a generation "is usually figured at about 30 years."[6] Then start counting at zero years with the "current" generation, beginning in 1941. Adding 30 years, brings us to 1971. Now the changes given for New York City are to occur sometime *in* the next generation of 30 years' duration, or anytime between 1971 and 2001. Thus, 2001 is the latest date by which the changes must occur for the earliest of the above two predictions to be fulfilled.

Catastrophes and Cataclysms

Uniformitarianism has been a long-standing doctrine of the Earth sciences. It asserts that throughout all the history of our planet, the Earth and its life have been shaped by the same slow processes that can be seen today wearing away mountains, causing sea-level to rise and fall, and replacing one species with another. Since no one has ever witnessed a sudden acceleration or deviation of any of the various processes, uniformitarians regard as outlandish any explanations of Earth history that involve sudden or unusual variations in natural processes.

In 1980, however, a team of scientists under the leadership of Walter and Luis Alvarez hypothesized that 65 million years ago, at the end of the Cretaceous, a huge comet or meteorite slammed into Earth with an impact 10,000 times the explosive force of all of the world's nuclear bombs, causing the dinosaurs to disappear. By 1993, over 2,500 papers and books had addressed various aspects of the Alvarez hypothesis, nearly all of them supportive. "In terms of its influence on future science, [William] Glen [a geophysicist and historian who has tracked the impact debate] ranks the impact hypothesis even above that of the plate tectonics revolution of the 1960s."[7] Today, many former uniformitarians have come to accept the doctrine of catastrophism.

A catastrophist caters to the notion that a few sudden, violent, short-lived, more or less worldwide events outside our present experience or knowledge of nature have greatly affected the physical conditions and inhabitants of Earth's surface. The Cretaceous impact event qualifies as a major catastrophe—or worse. It may have been a cataclysm. A cataclysm is "a momentous and violent [worldwide geophysical] event marked by overwhelming upheaval and demolition" *(Webster's New Collegiate Dictionary)*. Since 1980, scientists have documented at least four other times during the last 600 million years when other geophysical cataclysms have knocked life for a loop. The most severe of these killed more than 95 percent of all marine species.

Many of Cayce's prehistorical readings fit the catastrophic model of Earth changes. Consider the readings that say

• that "the deluge was not a myth (as many would have you believe)" (3653-1)

• that over the last 10,500,000 years many lands "have appeared and disappeared again and again" (5748-2), too quickly for today's uniformitarians and occurring in places like the Atlantic and Pacific Ocean basins where vertical crustal stability is currently thought to reign supreme

- that man-made destructive forces produced a great eruption which caused the portion of Atlantis now near the Sargasso Sea to go "into the depths" (364-4, p. 150)
- that at least two pole shifts occurred in the last ten million years (5249-1, p. 31, and 364-13, p. 30).

A strong catastrophic bent is found also in the readings' predictions of future Earth changes. We read, for example, that a pole shift is to occur, or to begin to occur, by the end of the present century (378-16, p. 23, and 826-8, p. 21). And all of the events in the many predictions given on pages 1-3 are also clearly catastrophic by the definition given above. They are the kinds of events that will occur either far too quickly or will be far too severe to fall under the doctrine of uniformitarianism. Now the readings don't say exactly *when* any of these events are to occur, but reading 364-8 (p. 235) indicates that "the changes will materially come" over the 1,000-year period of the first resurrection . . . [when] "He will walk and talk with men of every clime." And reading 5748-5 (p. 235) indicates that the Messiah will enter "in this period—1998." Thus, we may expect that all of the catastrophic events listed on pages 1-3 will take place between 1998 and 2998.

But reading 1602-3 (p. 23) says that "This [1998 change between the Piscean and the Aquarian age] is a gradual, not a cataclysmic activity in the experience of the Earth in this period." The implication here is that while individual catastrophic geologic events—like the destruction of Los Angeles, San Francisco, and New York—may occur over the next 1,000 years, humanity is not in for a cataclysm, like an event that would wipe out 95 percent of life on Earth.

POLE SHIFT

On pages 154 and 155, I postulated that a pole shift was initiated in 19,400 B. P. that took the North Pole from a position in northern Greenland to its present location. I said that this was due in part because men on Atlantis tuned the firestone too high and "produced what we would call a volcanic upheaval" (877-26). This upheaval propelled a large part of Atlantis "into the depths" (364-4), and forced the equator to move closer to the suddenly denser crustal volume comprised of the sunken subcontinent near the Sargasso Sea. The tendency of the equator to move closer to areas of higher relative density, resulting in a concomitant shift of the poles, is outlined on pages 37 and 38. Note also that additional areas of higher relative density had been developing for thousands of years prior to 19,400 B.P. beneath and within the European and North American ice sheets on adjacent continents to the northeast and northwest of Atlantis. The configuration of this distribution of excess ice mass, in a broad arc around the part of the Atlantean continent that sank, helped determine the path of pole motion hypothesized here. Sudden sinking of a large part of Atlantis was all that was needed to initiate the shift in the North Pole from its former position in northern Greenland to its present pole position. According to my interpretations of the readings (see Table A1, p. 254), initiation of this pole shift occurred around 19, 400 B.P.

Support for this inferred pole shift, although rather general, comes from the following reading that I discovered only recently.

Before that we find the entity was in that land now known ... as the Atlantean, during those days when there were the attempts of those to bring

quiet, to bring order out of chaos by the destructive forces that had made for the eruptions in the
land that had divided the lands and had changed
not only the temperate but to a more torrid region
by the shifting of the activities of the Earth itself.

884-1
April 9, 1935

Geoscientists have been trying for many decades to
explain the rapid onset of melting of the ice sheets in
Europe, Asia, and North America at the end of the last of
continental-scale glaciation around 20,000 years ago.
Here we have, as an explanation, man-induced volcanism and tectonism that caused a pole shift. A pole shift
that brought, quite rapidly geologically speaking, the
North American and European temperate regions that
lay to the south of the ice sheets, and on Atlantis also,
into "a more torrid region," causing the southern portions of the ice sheets to begin rapidly to melt.

The above reading suggests the following hypothetical scenario, which is best appreciated by reference to a
globe. Coincident with the sudden deep sinking of a subcontinental-sized block of the crust, called Atlantis, near
the Sargasso Sea, the equator began to move northward
along the 37°W meridian. It eventually moved a maximum of about 10 degrees of latitude. (Imagine the point
on the equator before the catastrophe as being somewhere between Recife and Salvador, Brazil. It then
moved to its present position 400 miles north of Mos-
soro, Brazil, in the Atlantic Ocean.) The North Pole also
moved a corresponding 10 degrees, along the same meridian, from northern Greenland to its present position.

We can only speculate on the length of time it would
have taken for a 10° pole shift to occur. Perhaps millen-

nia were required. One wonders if any confirmatory evidence—such as paleomagnetic data—might exist for the period 20,000 through, perhaps, 8,000 B. P.

Table A1 fills a void in the main part of the book. It lists concisely the various dates that I worked out that relate to the destruction of Atlantis, as well as readings' dates for the beginning and ending of construction of the Great Pyramid and the Library at Alexandria, Egypt. Note that it's sometimes difficult to distinguish between social upheavals and geophysical events, as in the following reading where "destruction to the land" is mentioned. Does "land" here mean physical land or "the country"? In this case, I would say the former, because "the use of those influences that brought destruction to the land" correlates well with other readings that describe similar events which took place around 19,400 B.P., when volcanic upheavals and sinking of part of Atlantis "into the depths" are mentioned.

> **Before that we find the entity was in the Atlantean land, when there were those activities that brought the first upheavals, and the use of those influences that brought destruction to the land.**
>
> **For the entity then was among those of the Law of One, but was persuaded by and with the leaders of the land to apply spiritual laws for MATERIAL gains.**
>
> **1292-1**
> **November 15, 1936**

Note that the dates in Table A1 are broadly similar to those listed by Edgar Evans Cayce in the book, *Mysteries of Atlantis Revisited*,[8] but that the 19,400 B.P. date is not mentioned in that book. To arrive at the 19,400 B.P. date, I assumed that the date of the final destruction of the last remaining piece of Atlantis was 11,900 B.P. Reading 288-1 says that this final destruction happened "nearly ten thousand years before the Prince of Peace came," or about 11,900 B.P. To this date I added 7,500 years because

Table A1. Dates of Important Geophysical Events in the History of Atlantis According to the Edgar Cayce Readings and the Author's Interpretations.

Date in Years Before the Present (B.P.)	Geophysical Event(s)	Reading Nos.
52,720	Beginning of pole shift from unknown position on Earth to northern Greenland	262-39
30,000	"Second period of disturbance" when "there were small channels through many of the lands"	470-22
24,000	Period of the "second of the eruptions"	364-6
19,400	Man-induced eruption causes the portion of Atlantis near the Sargasso Sea to go "into the depths." Volcano-like upheaval separates Atlantis into five islands and induces a pole shift causing ice sheets to melt in Europe, Asia, and North America	364-11, 364-4. 877-26, 440-5, 1291-1 (?)
12,700 to 11,900	" . . . WASTING away in the mountains, then into the valleys, then into the sea itself, and the fast disintegration of the lands" and then, the last island of Poseidia sinks below the waves.	364-4, 288-1
Related Cultural Events		
12,488 to 12,388	Construction of the Great Pyramid in Egypt	5748-6
12,300	Establishment of the Library at Alexandria	315-4

364-11 says that the date of the first destruction (as described on p. 151) was 7,500 years before the final destruction. Thus, the date for the "first destruction" is 19,400 B.P. This date correlates well with the approximate time that geologists have determined for the beginning of rapid melting of the last ice sheets.

Two final notes on pole shift. Much was made on page 29 of the 1996 announcement that Earth's inner core of solid iron is spinning freely within the molten outer core and that the axis of rotation of the inner core is tilted a few degrees with respect to Earth's north-south axis of rotation. This information was related to reading 5748-6 that says that great upheavals would take place in the interior of the Earth in 1936, due to "the shifting of same by the differentiation in the axis as respecting the positions from the Polaris center."

We now have a 1997 report[9] by two of the scientists who found that Earth's inner core of solid iron is rotating eastwards at about one degree per year faster than the mantle and that its axis of rotation is tilted about 10° from Earth's pole of rotation. P. Richards and X. Song have recently made seismological analyses that directly support numerical modeling efforts predicting that the inner core is driven to rotate eastwards by jets at the base of the fluid core. As postulated on page 30 of this book, there may be at least some scientific basis for the psychically derived statements about upheavals in the interior of the Earth, starting in 1936 and leading to rising thermal plumes in the mantle that will be partially or fully responsible for a shifting of the Earth's poles, starting in 1998.

There is new research that suggests that the hypothesized pole shift of 19,400 B. P. could have resulted *in a displacement of Earth's spin axis,* a mechanism of pole shift favored in the readings (see p. 30). Writing in the

journal *Nature*, B. Steinberger and R. J. O'Connell researched those dynamical arguments that show "that
polar wander can arise from the redistribution of mass
in a plastic deformable Earth, the rate of which depends
on both the rate of mass redistribution and the rate at
which the Earth's rotational bulge can adjust to the
changing rotational axis."[10] The authors used a viscosity
structure obtained by geoid modeling, a mantle flow
field consistent with known tomographic anomalies,
and lithospheric motions to calculate the movement of
mantle density differences and corresponding changes
in the geoid during the Cenozoic era. They showed that
Earth's rotation axis *will* "follow closely any imposed
changes of the axis of maximum non-hydrostatic moment of inertia"; that is, the Earth's rotational axis can
shift due to sudden imposition of a new distribution of
mass of sufficient magnitude.

What is the relationship of the previous information
to readings-style pole shifts? Assuming that the upheavals in the interior of the Earth in 1936 did cause plumes
of hot mantle material to rise toward Earth's surface,
then large-scale mantle density differences have been
forming ever since. And by 1998, these density differences may be great enough to cause the beginning of a
shift in Earth's poles. This shift will be the cause of, or
coincident with, upheavals in the Arctic and Antarctic,
the appearance of dry land off the Caribbean Sea,
breakup of the American West, and all of the other
seismotectonic events predicted in the readings.

IS AMERICA VULNERABLE?
THE WESTERN STATES
Breakup of the
Western Portion of America

The following lines from reading 3976-15 suggest possible catastrophes immediately ahead:

As to the material changes that are to be as an omen, as a sign to those that this [return of John the Beloved] is shortly to come to pass—as has been given of old, the sun will be darkened and the Earth shall be broken up in divers places ... [and] ... As to the changes physical again: The Earth will be broken up in the western portion of America.

The breaking-up mentioned in the last line above may be explained by (1) a 1936 reading on "upheavals in the interior of the Earth" (5748-6) and (2) new information on the deeper geologic structure of America between the Rockies and the West Coast (Fig. A1, p. 260).

R. A. Kerr, in a recent review article in *Science* (March 14, 1997, p. 1564), says, "New geophysical research shows that the lofty peaks and plateaus of the American West are buoyed up not by the continental crust alone, but also by deeper forces from Earth's mantle." He goes on to state that the West, floating on hot buoyant mantle beneath, is slowly spreading outward. Because it is hemmed in on most sides by rigid lithospheric plates, it may slowly "be heading for the Pacific Northwest, where plate motions open a small escape hatch."

Now if the speculation on p. 13 is true, that in 1936 "upheavals in the interior of the Earth" (5748-6) led to rising mantle plumes that will produce changes at the Earth's surface starting in 1998 (p. 29), an area that could be expected to be among the first to be influenced by such mantle motions would be the western states. The land there is sitting on a pool of hot mantle without the protection of thickened crustal roots formerly thought to underlie the topographic heights of the West. This makes the ground surface of the western states far more subject to deeper mantle motions than if it were underlain by the relatively cool and dense mantle that lies just beneath most continents. And recent work with seismic tomography contributes to a growing body of evidence for *mantlewide* convective flow.[11] Thus, hot rising mantle from upheavals in the interior of the Earth in 1936 could play havoc with the western U.S., starting in 1998.

Among the areas in the West that reading 270-35 (p. 68) saw as reacting to accelerating mantle motions were **"the southern coast of California—and the areas between Salt Lake and the southern portions of Nevada . . . "** Recent geophysical measurements that permit analysis of slowly changing distances between points on the surface show that the Colorado Plateau is hardly moving. But at Ely, in eastern Nevada, the land is moving westward across the Basin and Range Province (Fig. A1), "sliding from the heights supported by the hot, buoyant mantle," according to Kerr's review. In Utah, the "areas between Salt Lake and the southern portions of Nevada" lie roughly along the boundary between the Colorado Plateau and the Basin and Range Province. This boundary now appears to be a zone of crustal extension and, if suddenly subjected to accelerated mantle motions, the crust there could experience catastrophic earthquakes and faulting. That is, this part of "the western portion of America" would be "broken up."

But how do we explain breaking up of the southern coast of California by this new model of the mantle and crust of the American West? According to Kerr's research review, the push of land sliding westward from the heights across Nevada's Basin and Range is pushing the crust into the Sierra Nevada and "crunching up the crust along the San Andreas." That's the reverse of the conventional view, which holds that it's the Pacific Plate pushing eastward that shoves up the Coast Ranges (Fig. A1) along the length of California. However, the stress on the San Andreas and associated faults in southern California would be exacerbated by accelerated mantle movements beneath the western states, leading to breaking up by a crustal-compression mechanism.

The full paragraph of reading 270-35 indicates that we "may expect" significant earthquakes in the western areas just discussed.

> **If there are the greater activities in the Vesuvius, or Pelée, then the southern coast of California—and the areas between Salt Lake and the southern portions of Nevada—may expect, within the three months following same, an inundation by the earthquakes. But these, as we find, are to be more in the Southern than in the Northern Hemisphere.**
> **January 21, 1936**

This paragraph of reading 270-35 was not making a prediction for 1936, as has been suggested elsewhere.[2] The paragraph was simply continuing an explanation of the causes of earthquakes, the previous paragraph of which is to be found at the bottom of page 65 and top of page 66. Thus, the paragraph in bold type immediately above is saying that when the conditions described are met—greater activities in Mts. Vesuvius or Pelée—inundations by earthquakes will occur in the specified areas of the American West. But these "inundations" will be

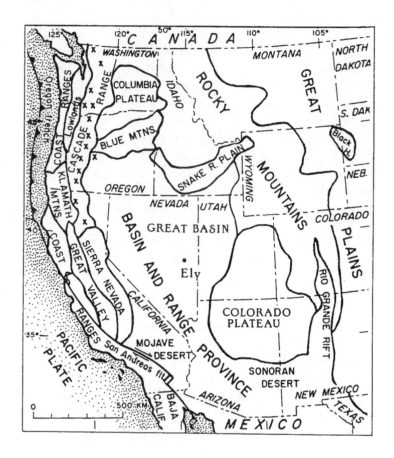

Figure A1
Physiographic provinces of the western United States. Xs indicate major young volcanoes above the subducted Gorda and
Juan de Fuca lithospheric plates. (See Fig. 9, p. 80.) These
plates lie to the west of the Oregon trench, plate-convergence
zone, where they begin underthrusting the North American
plate to the east.

more pronounced in the Southern Hemisphere. Presumably, after 1936, significant activities in either of the explosive volcanoes mentioned will signal the beginning of mantle and lithospheric-plate movements on a global scale. Although intermittent eruptive activity occurred at Vesuvius between 1913 and 1944, that occurring after 1936 was apparently not sufficient to signal the consequences predicted in the reading. The most recent explosive eruption of Vesuvius, in 1944, caused ash-cloud darkness at Bari, 200 km from the volcano. Apparently, "greater activities" at Vesuvius must exceed the 1944 level of activity. Perhaps ominously, seismicity at Vesuvius during 1995-'96—often a precursor to eruptive activity—was the highest in the past 50 years. Mt. Pelée, on the island of Martinique in the Caribbean, has been quiet since 1936.

Conventional interpretation of the phrase "an inundation by the earthquakes" has been that it is a figure of speech for "an overwhelming number of earthquakes." But a recent analysis of the earthquake hazard in the Salt Lake City area, by two seismologists at the University of Utah, suggests a different interpretation. These researchers modeled land deformation effects of a hypothetical, magnitude 7.2 quake that could be expected to occur on the south Weber segment of the north Salt Lake City segment of the Wasatch fault. (See p. 88 for the position of the Wasatch fault.) The seismologists put into their model a "scenario quake" consisting of the same ground-surface deformations that were observed in the M 7.5, 1959 Hebgen Lake, Montana, quake that occurred 270 miles north of Salt Lake City. They concluded:

> ... an unusual earthquake hazard [exists] ... due to possible flooding and *inundation* [emphasis added] by the Great Salt Lake accompanying large

> normal-faulting earthquakes on the Wasatch Fault
> ... Because of the close proximity of the Great Salt
> Lake to Salt Lake City, we demonstrate that this
> unappreciated hazard is very important, depend-
> ing upon the lake level and the location of the sce-
> nario earthquake.[12]

The changes implied by reading 270-35 above would involve far more than the isolated-earthquake, land-deformation, and flooding scenario modeled by the Utah scientists. Our new interpretation of reading 270-35 is that catastrophic earthquakes and flooding may occur in the land areas around Great Salt Lake and Utah Lake in Utah, around and below Lake Mead in far southern Nevada, and along those land areas fronting the Pacific Ocean in southern California. Such aqueous inundations could directly or indirectly affect population centers like Bringham City, Salt Lake City, Murray, Orem, and Provo in Utah, Las Vegas in Nevada, and the cities of coastal California from Santa Barbara south to San Diego.

Thus, we now have two ways of interpreting the "inundation-by-the-earthquakes" language of reading 270-35. And the ambiguity of the phraseology used seems to have allowed science to catch up to the reading, here near the end of the 40-year period (1958 to 1998) said to be prelude to catastrophic Earth changes between 1998 and 2998. As if to bolster our interpretation at this critical time, there is new global-scale seismic-tomography evidence[11] for deep mantle circulation of unprecedented proportion.

There's no question that the language of the readings can be quite challenging at times, as outlined in the examples above. But in continually trying to understand, we may in fact get understanding, often in unexpected ways. Here, recent scientific findings are accommodated by seemingly obscure phrases in the readings, many years after Cayce gave them. Were certain of the readings intentionally constructed this way? If so, we may have insight here from the promise that, "In 1998 we may find a great deal of the activities as have been wrought by the gradual changes that are coming about . . . [and] . . . As has been indicated, we will begin to understand fully in '98" (1602-3).

Los Angeles and
Coastal Southern California

What are the chances of "an [aqueous] inundation of the southern coast of California" due to earthquakes, as just discussed? Have there been any earthquakes and associated tsunamis (seismic sea waves) there before? Yes, indeed.

But first, just what are tsunamis? (The word is Japanese for *tsu* harbor + *nami* wave). A tsunami (pronounced tsoo-nah-mee) is a series of waves, generated in a body of water by an impulsive disturbance that vertically displaces ocean water. Sea-floor movements accompanying earthquakes are the most common cause of tsunamis. But landslides into the sea, submarine landslides, explosive submarine volcanic eruptions, and even meteorite impacts can generate tsunamis. Locally generated tsunamis are the greatest threat to U.S. coastlines, but tsunamis originating at a distance are also a constant threat. The U.S. has suffered major damage from tsunamis originating in Chile, Japan, Russia, and Alaska. Eventually, tsunamis will strike all U.S. Pacific Ocean coastlines.

Because reading 270-35 above refers to tsunamis generated by local earthquakes, we will cite details for sev-

eral here to give the reader an idea of historical numbers
and effects. As covered in *Tsunamis Affecting the West
Coast of the United States: 1806-1992*,[13] the Santa Barbara
tsunami of December 21, 1812, was most probably
caused by a submarine landslide in the Santa Barbara
basin. This landslide was believed due to an earthquake
estimated at M 7.7 and originating on the San Andreas
fault. The maximum wave height was about 15 feet or so,
and is said to have reached half a mile inland. Submarine landslide tsunamis in California typically have
waves with a maximum amplitude of 10 feet and affect a
very restricted area. The tsunami of November 4, 1927,
was such a one and affected about 35 miles of shoreline.[13] It was recorded as a six-foot wave at Surf, just north
of Pt. Arguello near Santa Barbara, and as a five-foot
surge at Port San Luis. Near Los Angeles, on July 10, 1855,
four earthquakes generated a "probable submarine
landslide and local tsunami."[13] On May 27, 1862, an M
5.9 earthquake at San Diego caused the only local tsunami observed there. On August 30, 1930, an M 5.2 earthquake caused a 20-foot wave at Santa Monica, Venice,
and Redondo Beach: "Probably a submarine landslide
source given the low magnitude of the earthquake and
localized effect. Sixteen people were rescued from the
surf. One drowned at Redondo Beach."[13] These few representative records show that local tsunami inundations
"by [the] earthquakes" are quite possible for the southern coast of California.

In all, some 21 locally generated tsunamis of various
destructive power have been documented for the southern California coast between 1806 and 1992.[13] Although
this coast *seems* safe from locally generated tsunamis
relative to the coasts of, say, Japan or Alaska, the dangers
from large tsunamis cannot be ignored. D. S. McCul-
lough, in a U.S. Geological Survey report on tsunamis

along the Pacific coast, says that "a preliminary appraisal of tı potential for locally generated tsunamis suggests that ı ve run-up heights as great as four to six meters," or 13 ı 20 feet, could be caused by seabed movements due to earthquakes. Such waves are not in the same league with the giant tsunamis that hit Hawaii in 1946 and 1975, but even a five- or six-foot tsunami can cause widespread damage in a harbor or along a heavily settled stretch of coastline.

One tsunami-prone area in southern California is located off Santa Barbara where the offshore area has created an odd set of conditions with the potential for making large seismic sea waves. Quoting a report on southern California's tsunami potential, D. Ritchie says in his book *Superquake!* (Crown, N.Y., 1988), "Surface fault rupture accompanied by sea-floor displacement is a distinct possibility beneath the Santa Barbara Channel. To put it another way, a big quake on land could set off another quake under the sea and thus send tsunamis rolling toward the California shore."[13] Support for this view is found in J. Deng and L. Sykes's recent study of the 200-year evolution of the crustal stress field in southern California (*Jour. Geophysical Research*, v. 102, B5, May 10, 1997). The authors write: "Future great earthquakes along the San Andreas fault, especially if the San Bernardino and Coachella Valley segments rupture together, can trigger moderate to large earthquakes in the Transverse Ranges, as appears to have happened in the Santa Barbara earthquake that occurred 13 days after the great San Andreas shock of 1812."

We note in passing that another kind of inundation that could occur would be due to flooding by reservoir waters released when earthquakes cause dams to fail. Several stream-dammed reservoirs are located in the coastal zone of southern California.

If the ultimate cause of the tsunamis mentioned in our speculation on the meaning of 270-35 above is an acceleration of lithospheric plate movements incident to pole shift, there is one last, far-out possibility to consider. And it ties into the last sentence of the reading which indicates that the effects of the inundations by the earthquakes will be felt more in the Southern Hemisphere. This understanding of the phrase "inundation by the earthquakes" would relate to tsunamis generated from distant sources. If "the greater portion of Japan [goes quickly] into the sea," or if a huge piece of Kilauea volcano, Hawaii, suddenly slides into the Pacific, highly energetic tsunamis would be generated that would strongly impact the southern California shoreline and those of Central and South America. Indeed, the coast of southern California has experienced no less than 44 tsunamis generated by earthquakes or submarine landslides beyond the U.S. Pacific coast (excluding Alaska) between 1806 and 1992. These waves have traveled from Japan, coastal Alaska, Chile, the Kuril Islands, the Java Sea, El Salvador, and other places. Many of the 44 tsunamis of distant origin that hit the coast of southern California were recorded only as minor excursions on tide-gage records. But consider what could happen as accelerating mantle motions cause earthquakes and landslides at currently erupting volcanoes like Kilauea, Hawaii. (The seismic hazard for the southern portion of Hawaii rivals that of the area around the San Andreas fault in California, according to F. Klein, seismologist with the U. S. Geological Survey. And geophysicists in early 1997 were measuring progressive seaward displacements of the land portion of Kilauea's southeastern flank.)

First, note that the tsunami of December 21, 1812, mentioned above traveled all the way from Santa Barbara, California, to Hookena, Kona, Hawaii, where the wave run-up height was estimated at between six and fourteen feet.[13] Now let's reverse the California-to-Hawaii tsunami travel direction. Suppose that a giant submarine landslide occurs on the southeastern flank of Kilauea and generates a large tsunami. The coastlines of Peru and Chile would receive the brunt of such a tsunami, but the coast of southern California would also be affected. Recall the words of 270-35, *"But these [inundations by the earthquakes] are to be more in the Southern than in the Northern Hemisphere."* In their paper, "Giant Hawaiian Landslides" (*Annual Reviews of Earth and Planetary Science,* 1994), J. Moore, and others, document dozens of major landslides that have recently been discovered on the flanks of the Hawaiian Ridge. They are among the largest on Earth, attaining lengths greater than 200 km and volumes of several thousand cubic kilometers. The authors write that rapid movement of some of the submarine debris avalanches "is indicated by the fact that some have moved uphill for tens of kilometers, and are believed to have produced major tsunamis."

Now just to cover all possibilities in our interpretation of reading 270-35 above, let's think about the spelling of the word "Pelée." Was it left to Gladys Davis, Edgar Cayce's stenographer for the reading, to decide on the spelling of this word? Or did Cayce, while giving the reading, spell it for her? We don't know for sure, but we do know that most of the time it was left to Miss Davis to come up with the spelling of unfamiliar words in Cayce's readings. But what if the word was really "Pele," a difference of only one letter "e"? In Polynesian mythology, Pele was the fire goddess who inhabited the crater of Kilauea volcano, causing it to erupt during her jealous rages. Thus, the phrase, *"greater activities in [the Vesuvius] or*

Pelée" might really be referring to "the domain of Pele," on Kilauea volcano on the Big Island of Hawaii.

I don't think that this last interpretation is the best, but maybe it doesn't matter because once mantle movements incident to pole shift begin to affect sensitive calderas, and volcanoes like Vesuvius, and Torrid-Zone Pelée and Kilauea, "[aqueous] inundations by the earthquakes" will become almost commonplace. The Hawaiian Volcano Observatory calls Kilauea "the nation's deadliest volcano," pointing out—among several other deadly events documented by the Observatory—that 46 people perished along Kilauea's coastline in 1868, when a huge earthquake caused widespread shaking, coastal subsidence, and a tsunami that destroyed several villages.

SOCIETAL TRENDS AND AMERICA'S FUTURE

More reasons are given in this Addendum to expect an "awful [karmic] woe" to come upon America and the rest of the world, as mentioned by Phylos (p. 108) in his visions of the end of the present age. Phylos asserted that, for placing violence, lust, and greed in command of his life, "man is his own judge and executioner." And "Soon millions of trained soldiers will turn upon the visible representatives, the wealthy and worldly prosperous . . . [later] . . . they will break up into lawless bands bent on satisfying Ishmaelitish tendencies, each self-server's hand weaponed against their fellow creatures." (p. 109)

We addressed this last quote of Phylos (p. 111) by citing as an example of the coming anarchy the empowerment of private armies and the collapse of social order in West Africa, with special reference to Sierra Leone. In late May 1997, the U. S. Navy rescued 330 Americans and 600 foreign civilians from Freetown while "shabbily dressed, gun-toting [army] mutineers burst into homes and businesses, taking what they could, and often burning them when there was nothing to be grabbed" (Asssociated Press). To this we now add Zaire's rebel takeover of the Mobutu government in which government forces joined the rebel side, a lawless Liberia,[14] and the armed incursions by veterans of Rwanda's defeated army into Uganda, Burundi, and Angola. And now on our watch list for Phylos-type collapses of the social order are Russia and Mexico.

Russian Defense Minister Igor Rodinov met with his top advisors on March 14, 1997, to discuss the rapid decline in morale of Russia's officer corps. "More than 500 Russian Army officers committed suicide in 1996, and 20% submitted their resignation. The Russian military is getting increasingly worried that they have destroyed their military by housing them in such miserable conditions and then not being able to pay them . . . Without officers to lead them, the army is useless—but even more frightening is the possibility that the remaining officers would lead their troops to revolt over their horrible treatment since independence."[15]

In Mexico, where one cannot tell where the government ends and the drug cartels begin, *Time* magazine reported on March 10, 1997, that "One top official describes the Mexican police force as 'the most corrupt institution in the hemisphere' and denounces the ruling Institutional party (PRI) for promoting and tolerating corruption. 'The PRI is a dictatorship on the verge of collapse,' says this official. 'As it goes, it is confronting Mexico with quite fundamental questions about the rule of law in that society.'"

Through 1996, the only institution left in Mexico that was seemingly free of drug scandals, and that commanded the respect of almost everyone, was the Mexican Army. Then, in early 1997, the Army general in charge of Mexico's Institute to Combat Drugs, Jesus Gutierrez Rebollo, was found to have been on the payroll of the country's number-one drug lord, Amado Carrillo Fuentes, for several years. "According to a top U. S. military intelligence official, the only glue holding Mexico together is the Army... Once the Mexican people believe their Army is just as corrupt as their government, the revolts begin in earnest."[16] Incipient armed rebellions are already occurring in a dozen or so Mexican states, such as Guerrero, a state of great poverty beyond the foothills that ring the luxury tourist resort of Acapulco. Once the Mexican people see that the Army is incompetent to contain the revolts—look at their miserable performance against the Zapatistas in Chiapas— the prospect of full-blown civil war is all too likely. And defectors from the Army may be expected to join the rebels. Mexico's last Civil War, from 1910 to 1920, took some ten million lives.[16]

As for remarks on page 111 about America being "the world's leading merchant of death," the situation has gotten worse. "The latest example is the Clinton Administration's decision to allow American companies to submit bids to sell F-16 fighters to Chile, a country that has no need for advanced warplanes. The sale would start a potentially dangerous and destabilizing arms race among Latin American countries that can ill afford it."[17] As much as 85 percent of the arms Washington sells to developing nations strengthen repressive and corrupt militaries. Since 1993, the U.S. has stopped paying even lip service to the notion of caps on regional arms sales. At present, the new world arms market involves a fierce competition between Russia and China vying with America and Western Europe to capture market share— whatever the cost—especially in the Persian Gulf region.

Since 1990, the Gulf states have signed contracts for $36 billion in American arms, or 32 percent of the $110.8 billion in U.S. arms exports worldwide over the same period, according to an analysis of Defense Department figures by the Washington-based Arms Control Association.[18] The U.S. Government employs 6,500 people to promote and service arms sales, and tax breaks to arms manufacturers and low-interest loans to client countries to buy weapons altogether amount to $7.5 billion a year.[17] We think again about the reading that says, "Man's answer to everything has been power!" As the major global purveyor of high-tech weapons of violence, cannot America expect to reap a whirlwind of violence herself, including, perhaps even, the violence of Earth changes?

Finally, we must not overlook the culpability of America in Zaire's descent into savagery in Africa, and Guatemala's descent in Latin America. As Anthony Lewis has asked,[19] "What responsibility should Americans feel when our Government, for policy reasons, imposes on another country a regime that reduces it to misery?" Lewis reminds us that the C.I.A. backed Colonel Mobutu when he conducted a coup in 1965 and made himself president. Over the next 30 years the U.S. supplied more than $1.5 billion in economic and military aid to Zaire, much of which Mobutu spirited away to Swiss bank accounts. Zaire was left to decay into chaos.

"Zaire is not the only example of U.S. Government's using the C.I.A. to reshape the politics of another country, indifferent to the cruelties imposed on the people of that country. Guatemala is another, perhaps equally outrageous from the viewpoint of international law and human values," Lewis says. He goes on to describe how the C.I.A., in 1954, overthrew the left-wing government of President Arbenez. The Arbenez regime, however, was one of the few that had ever been freely elected in that country. Then, in the 1980s, "Guatemalan military forces carried out a relentless campaign against rebels among the Indians who make up a majority of the country's population. The military strategy was to destroy villages deemed sympathetic to the rebels. In the end, 140,000 people were killed or missing and one million displaced from their villages."[19] Will America suffer no retribution for these outrages against humanity?

WILL JAPAN BE SUBMERGED?

Speculations as to the credibility of reading 3976-15, to the effect that "The greater portion of Japan must go into the sea," were based both on recent geophysical evidence (pp. 126-130) and the metaphysical effects of negative societal trends (pp. 220-223) in Japan. We said that northern Honshu and all of Hokkaido might constitute "the greater portion of Japan" mentioned in the reading, and that near the center of this region was Rokkasho (see p. 127) on the northeasternmost coast of Honshu. It is here, in the Tohoko District, that a large Japanese plutonium-recovery complex is being built. Is there any evidence for recent subsidence in this general area? Yes, there is.

Recent results of a study[20] of vertical crustal movements in the Tohoku District of Japan identify a trend that coincides with our geophysical speculations. This district encompasses all of northern Honshu above a line connecting, roughly Iwaki and Niigata, on opposite coasts. Although some small areas within the district have shown net crustal uplift during the 1966-1995 study interval, the authors' primary conclusion is that "the most fundamental trend is extensive subsidence along the Pacific coast." This conforms to the period 1958-1998 in which signs of coming Earth changes are to appear.

SHAKY LATIN AMERICA

Prior to 1994, Popocatepetl, the giant volcano mentioned on p. 135, had not been active in the 1958-1998 period. But on July 1, 1997, it staged its biggest eruption since 1925, coating Mexico City with ash. The tsunami situation in the Lima-Callao area of Peru is relatively representative of the entire Pacific coast of South America where a subduction zone lies offshore. Most Peruvian tsunamis are produced by local submarine earthquakes at shallow subduction depths between the Nazca and South American plates. (See Fig. 5, p. 56.) Callao has been destroyed at least three times, the most significant tsunami being that of October 28, 1746. That tsunami produced a run-up wave about 75 feet high, "leaving only rubble where the town once stood." Numerous large earthquakes have occurred in Peru, some seven alone, since 1951, ranging from M 5.3 to 7.8. And thus, when the mantle and lithospheric plates start to move more rapidly in 1998, we can expect a like increase in such geophysical events all along the South American Pacific rim.

HUMANITY'S RESPONSE TO EARTH CHANGES
Nature Taking a Hand

According to Rolling Thunder (p. 210), some of the things that are beginning to happen in the way of natural disasters are Earth's attempt to throw off some of its "sicknesses." The sicknesses are due to over-exploitation and unbalancing of nature by greedy humankind, and to the poisoning of air, water, and soil by man-made chemicals. Heightened awareness of the magnitude of two significant unbalancings of nature is found in (1) dismaying new figures for the destruction of the Amazon rain forest by fires and logging and (2) the toxic legacy of Britian's oil boom. New data show that deforestation rose 34 percent, from the 1990-1991 Amazon burning season to 1994,[21] contradicting Brazil's claims that stricter environmental laws had slowed destruction of the Amazon rain forest. And in the North Sea off Britian, more than a million tons of oily waste lying on the bottom is polluting wide areas with oil and heavy metals, such as cadmium and mercury, killing marine organisms over wide areas.[22]

NATURAL CONSEQUENCE
OF GOOD JUDGMENTS
Moralistic Reasons for Earth Changes

Under this heading, we raised the specter of a hypothetical Earth change taking place—for moralistic reasons—as the result of activities at Lawrence Livermore National Laboratory (LLNL), near the Hayward and Calaveras faults east of San Francisco. (See pp. 218-219.) What's new since then? LLNL is part of a $4-billion-a-year federal program at the nation's three weapons labs to make sure that the country's nuclear arsenal remains safe and ready to use if needed. The crown jewel of this "stockpile-stewardship program" is the $2.2 billion National Ignition Facility. Ground-breaking for the NIF was May 30, 1997. The complex will be about the size of the Rose Bowl and contain a cluster of high-powered lasers that will focus in unison on a tiny hydrogen target, "blasting it with so much energy that it will undergo a miniature thermonuclear explosion."[23] The hydrogen will be heated instantaneously to a temperature hotter than the sun's surface, causing it to fuse into helium in a burst of pure fusion energy.

Once again, we have souls involved in the power game for, "Another important goal for NIF [is] to develop a cadre of scientists who know how to design new [nuclear] weapons before all the old hands retire or leave."[23] And meanwhile, weapons scientists at Livermore, Los Alamos, and Sandia National Laboratories are working on small, pure-hydrogen bombs that can be easily transported. When will this type of national program end? When America develops a new moral consciousness.

There is another, chilling dimension to the use of powerful lasers in the LLNL research program, as suggested in reading 364-11 immediately below. As an introduction to the reading, note that lasers are made up of several basic components, including the active medium, which may consist of atoms of a gas or ions in a crystal. Another component is some method for introducing energy into the active medium, such as a flash lamp, or other energetic beam. The reading deals with the positive and negative aspects of psychic science as it was practiced on Atlantis before, and then during, the first man-made destruction of that subcontinent.

> Even as with the use of those sources of [psychic] information, the abilities to become a portion of those elements that were the creative forces OF the compounds or elements within the universal forces, at that period brought about those forces that made for destruction of the land itself, in the attempt to draw that as was in man then back TO the knowledge; and these brought about those destructive forces ... *in gases, with that called the death ray* [emphasis added], that brought from the bowels of the Earth itself—when turned into the sources of supply—those destructions to portions of the land. Man has ever (even as then) when in distress, either mental, spiritual OR physical, sought to know his association, his connection,

with the divine forces that brought the worlds into being. As these are sought, so does the promise hold true—or that given man from the beginning, "Will ye be my children, I will be thy God!" "Ye turn your face from me, my face is turned from thee," and *those things ye have builded in thine own endeavor to make manifest thine own powers* [emphasis added] bring those certain destructions in the lives of individuals in the present, even as in those first experiences [on Atlantis] . . .

Thus, as American scientists and engineers at LLNL and the other weapons laboratories in the American West seek to manifest their own powers with laser-based experiments that are perhaps an echo of those on Atlantis, would it be any wonder that they might "bring those certain destructions in the lives of individuals in the present"? Are the predicted Earth changes in the western states going to be induced by amoral weapons research that has been going on there for 50-plus years? Should not the western politicians who vote the funding for weapons research listen to Dr. Hans Bethe, a primary architect of the first atomic bomb? In April 1997, he wrote to President Clinton as follows: "The time has come for our nation to declare that it is not working, in any way, to develop further weapons of mass destruction." Bethe pointed specifically to the danger of "pure-fusion weapons" (W.J. Broad, *New York Times*, May 27, 1997).

One asks the above questions while noting that some geoscientists are saying that the Hayward fault—lying roughly 15 miles west of Livermore—currently has the highest probability of all of the faults in the San Francisco Bay Area for the next big earthquake.[24] Should the M 7.5 quake described on p. 74 occur on the Hayward fault, an inescapable social upheaval would engulf the more than five million residents of the area as the rupture and its shaking severed the lifelines of power, fuel, transportation, water, sewage, and communication that stitch together the social fabric.

Reading 820-1 below, given in 1935, provides a past-life basis for why certain souls are either strongly attracted to, or abhorred by, laser-powered nuclear weapons technologies. It also provides an outline for a new perspective on life for those who are captivated by such research today.

Before that we find the entity was in the Atlantean land, during those periods when there began to be the rebellious forces that disputed those acts and laws pertaining to the communications with what is termed in the present as the unseen forces—or to those hierarchies that are given rule over activities in the various forms in the Earth.[25]

The entity then, in the name Al-Aar, was the ruler of those forces of the Law of One; and those activities that were raised against the entity's activities by the Belzebubs (?) make for an influence in the PHYSICAL forces of the body of the entity in the present. Not that which may not be arighted, if the entity—as then—be not turned aside; irrespective of what may attempt to divert or to alter the purposes from the whole manifestation of the spirit of truth and light and love. Once lost in this direction, again those greater shadows of those impending forces that the entity saw in those mighty upheavals from the destructive forces used for the people from the prisms' activities and from the fires that were started for the fires of the deeper

inferno that brought to the surface those destructive forces as from nature's storehouse itself. Yet
the entity may, applying those same tenents that
were held to in that period, make for the greater or
farther advancement in the present. Much might
and power are in thine inner self, my son. Keep
those inviolate, if ye would be directed in that
whereunto thou may indeed explore those fields of
service and activity that may bring not only plea-
sure, comfort, and ease but—most of all—to thy
fellow man the greater knowledge of the love of the
Divine that would free each soul, if it will but ac-
knowledge His presence as the motivative influence in all its associations and dealings with its
fellow man. Not for self, but less and less of self,
and more and more for the glory of Him that came
to his own and His own received Him not. But as
thou hast heard him, even as Al-Aar, as thou directed those in bringing for thy fellows in the varied lands the knowledge of the Law of One, so may
thou in thine activities in the present rise not only
to the greater cause of satisfying those innate
longings, but may rise to the power—not for self,
but for thy fellow man—such as to bring to thee
thine greater development, thine own illumina-
tion and thine own understanding. And in all thy
getting, get love—with the deep understanding.

Also under the heading "Moralistic Reasons for Earth
Changes" a brief discussion was offered (p. 217) of the
locations of nuclear weapons research and production
activities and their rather uncanny correlations with current areas of hazardous seismotectonic potential. To
amplify this point, note that in January 1997, the U. S.
Department of Energy released long-secret information

on the most serious continuing problems of health and safety posed by inadequate storage at 13 sites. These sites are where the department's stock of 250 tons of highly enriched uranium is kept. "The problems stem largely from the fact that many of the facilities were built during the 1950s and have faulty or inadequate fire-protection systems. A large blaze . . . could spread radioactive debris outside the Energy Department facilities, thus presenting a potential public-health hazard. The most vulnerable sites include Oak Ridge [Tennessee]; the Idaho Falls Laboratory , Idaho; Rocky Flats Environmental Technology Site near Denver; and Los Alamos National Laboratory at Los Alamos, N.M."[26]

A final point about the locations of nuclear weapons research and production that cries out for mention is related to America's new nuclear "gravity bomb," as presented in G. Mello's article, "New bomb, no mission" (*Bulletin of the Atomic Scientists*, May/June 1997). Mello writes:

> **The B61 "mod-11" gravity bomb is the first new nuclear capability added to the U. S. arsenal since 1989. It was developed and deployed secretly, without public or congressional debate, and in apparent contradiction to official domestic and international assurances that no new nuclear weapons were being developed in the United States.**
>
> **The [weapon's] "unique earth-penetrating characteristics and wide range of yields allow it to threaten otherwise indestructible targets from the air . . . "**

Mello goes on to say that perhaps most of the advocacy for the weapon came from the weapons labs, particularly Los Alamos National Laboratory (LANL) located northwest of Santa Fe, New Mexico. And just to the west of LANL lies the Valles caldera which may be entering a new cycle of activity. (See page 40 for a discussion of caldera eruptions.) There is evidence that this caldera may be the site of renewed magma generation (J. Wolff and J. Gardner, *Geology*, May 1995). Previous eruptions of the caldera produced lava domes and tuffs over a wide area, including Los Alamos. The front-end parts of the new bomb "are (or were)" made at the Y-12 plant at Oak Ridge National Laboratory just west of Knoxville, Tennessee, at the northern end of the Eastern Tennessee Seismic Zone (p. 104). The tail components were made at the U. S. Department of Energy's Kansas City, Missouri, plant. Although not located in the America's western states, Kansas City lies within a small area of elevated seismicity (Hadley and Devine, 1974). And it lies about 100 miles east of the portion of the Nemaha Uplift earthquake source zone (see Fig. 14, p. 91) that produced the two largest historical earthquakes in the area. These occurred in 1867 and 1978, although they were only of intensity VII (see Table 1, p. 96).

The only weapons facility that we have not so far correlated with known seismic source zones is the Pantex Plant near Amarillo, Texas. This installation was used for years to assemble nuclear bombs. It lies in the Ouachita-Wichita Mountains earthquake source zone (see Fig. 14, p. 91), and between two currently seismically quiet faults: the Meers fault (see p. 97) to the east and the Cheraw fault to the northwest, on the Piedmont of southeastern Colorado.

MORALISTIC REASONS FOR HYPOTHETICAL EARTH CHANGES IN JAPAN AND CHINA

In remarks on p. 221, it was noted that North Korea had accused Japan of pursuing nuclear weapons and had used Japan's program to justify its own. We now learn that a high-ranking defector from the North has said that N. Korea does have nuclear arms and is capable of "scorching" the South and Japan.[27] And Japan is justifying *its* planning for a nuclear weapons program by reference to perceived threats from China and North Korea. But, you say, Japan is not supposed to be engaged in nuclear weapons activities of any kind because Article 9 of their constitution promises that Japan will "forever renounce war," and never maintain "land, sea, and air forces as well as other war potential." However, almost immediately after Emperor Hirohito stamped his seal on the constitution in 1947, Japan, with U. S. approval, began ignoring Article 9 by quickly building up its military, euphemistically called their Self-Defense Forces.

Consider, also, the short report *A Nuclear Japan?* which asserts that "A secret program to produce nuclear weapons has developed all the components necessary for an entire nuclear arsenal, including stockpiled uranium and plutonium, all ready for quick assembly . . . [and] . . . key Japanese military and intelligence personnel are urging both the deployment of a fleet of nuclear-armed submarines and a layered anti-missile defense for the entire island chain."[28] Is the foregoing evidence of redeveloping "domination forces" in Japan? (See p. 222.)

As for societal trends and moralistic reasons for a hypothetical Earth change in China (p. 220), consider the article by Patrick E. Tyler, entitled "China Raises Nuclear Stakes on the Subcontinent." We learn that "China's assertiveness . . . reflects a stronger influence by the Chinese Army during the political transition now underway after nearly two decades of dominance by Deng Xioping."[29] The article goes on to outline recent Chinese initiatives in aiding Pakistan to develop a missile factory and related activities, supposedly to create a strategic balance on China's western flank. In addition, "China has been outfitting Iran with poison gas ingredients and equipment for at least five years" and it "deliberately deceived Washington officials when it claimed it was importing American machine tools for civilian purposes. Instead, it diverted them illegally to a missile factory."[30]

These activities can all be read as attempts by China to dominate its neighbors. And consider Georgie Anne Geyer's (*Washington Times*, May 30, 1997) column in which she says: "To be specific, China wants to *dominate* [emphasis added] the 32,000 square miles of water in the South China Sea, thus giving it control over the oil-rich Spratly Islands (also claimed by Taiwan, the Philippines, Malaysia, and Vietnam). And since China also has disputes with Japan, Taiwan, South and North Korea, the Philippines, Vietnam, Malaysia, and Brunei, there is no lack of reason for conflict."

Again, Earth changes to stop the nuclear madness may be preferable to global nuclear war. (See p. 225.) And how close may we be to nuclear disasters? According to James T. Hackett, "Russia might soon reach a threshold beyond which its rockets and nuclear systems cannot be controlled," due to the reduced reliability of the command and control of Russia's nuclear forces.[31] Col. Bykov, a retired Russian colonel who spent much of his 33-year military career in the Strategic Rocket Forces, "warns that an accidental nuclear launch at the U.S. could happen at any time, in a matter of seconds." [He] "describes one incident in which a missile control duty officer was close to madness and had to be hospitalized." He describes another case in which a "smart aleck in a missile regiment figured out how to bring the launch contact points together, making it possible to launch a missile without the launch code."[31] We ask again, can humankind in its present state handle nuclear energy in all of its many forms? Have we gone too far with our linear-thinking, invention-producing capabilities? Is it not time to pay more attention to holistic, intuitive forms of mentation?

A rational solution to our problems is given in *The Axemaker's Gift*,[32] by James Burke and Robert Ornstein. The book details how we have through history used our minds in a way that has led to both our greatest successes and our largest problems. It "ends with a provocative suggestion about how modern information technologies may, paradoxically, embody both the ultimate fruits of this approach and the seeds of its transcendence."[33] The work details how, at each major step of innovation, from the first stone axe to the supercomputers of the modern world, those few with the capacity for sequential analysis (the axemakers) generated technologies that gave them the power to control and shape the rest of their community. The book explains how the culture we live in—based as it is on the sequential influence of language on thought, and operating according to the rationalist rules of Greek philosophy and reductionist practice, has wielded great power.

It has given us the wonders of the modern world on a plate. But it has also fostered beliefs that have tied us to centralized institutions and powerful individuals for centuries, which we must shuck off if we are to adapt to the world we've made: that unabated extraction of planetary resources is possible, that the most valuable members of society are specialists, that people cannot survive without leaders, that the body is mechanistic and can only be healed with knives and drugs, that there is only one superior truth, that the only important human abilities lie in the sequential and analytic mode of thought ... [32]

During this long period of implementation of rationalistic and reductionist activity, the other, older kinds of knowledge born of intuition and the brain's multiple nonverbal talents were largely ignored. Now, the authors say, the cumulative effects of axemaker technology have brought us to the point where it is possible and imperative for our survival to bring back into use those ancient forms of knowledge still resident in the non-axemaker cultures of the modern world.[33] Rolling Thunder (p. 210) comes to mind here.

Without meaning to slight the authors of *The Axemaker's Gift* in the least—for they have written a brilliant book—I wonder if they have ever encountered the Edgar Cayce readings. In particular, I think of the reading that says that while we all may have different ideas, all can hold the same ideal. (See pp. 198-200.) And there's, "Know thy ideal, and live to that. For, each soul must give account for its own self." (2803-2) But idealism itself can be tricky, as Idries Shah has so ably pointed out on page 213 of his book,[34] *The Commanding Self*. He writes, for example, that so-called idealists who want things done that are not solutions at all "are not idealists, but the destroyers of good . . . [and that] . . . Idealism should never exclude a desire to know the truth. If it does, it will destroy the much more valuable thing of which idealism is only a manifestation."[35]

I think that the concept of working with ideals needs much more emphasis in the world today. The Cayce readings have many useful things to say about the formulation and application of ideals. And the question as to *why* anyone ought to consider application of positive ideals in his/her life—as a technology for ultimate reunification with one's creator—is explained rather well by a Cayce reading on the subject of reincarnation.[36] This is important for:

> The first cause was, that the created would be the companion for the Creator; that it, the creature, would—through its manifestations in the activity of that given unto the creature—show itself to be not only worthy of, but companionable to, the Creator.

See note 36, p. 319, for the complete text of this important reading. It describes various aspects of God's desire for companionship, and of reincarnation and related soul development.

We leave our update on the doings of the nuclear weapons axemakers and our speculations as to their relationships to the Earth changes predicted in the Cayce readings. Will our physical, mental, and spiritual environments begin obviously to be changed by the end of 1998? By the year 2000? We have only to watch and to wait.

A NEW AGE, A NEW UNDERSTANDING
Earth Changes Map for the United States and Southern Canada

The map shown at the end of this addendum is the author's interpretation of what would happen to the United States and extreme southern Canada if all of the predictions in the readings were to come true.

Please note that I do not see America, as

> ... the world, as related to the Earth and its position in the universe, being damned irrespective of what souls do about same—taking little or no account of the words, the promises, yea the activities of Him [for] He manifested in the Earth that as would bring to the seeker an awareness of the con-

structive influence of same. (See paragraph 14, page 231.)

The map, "Earth Changes Predicted for the Contiguous United States and Southern Canada—1998 and Beyond," is approximate, speculative, and for illustrative purposes only. (**Note:** Neither the publisher nor the author make any warranties concerning any map in this book as constituting advice to readers and no maps in this book are intended to represent a source for life-altering decisions.)

Drawing boundaries and designating geophysical potentials on a map implies that the map-maker has a significant knowledge of the information conveyed. But this is not the case for the Earth changes map. For example, Virginia Beach, Virginia, "or the area" is referred to in one reading as a "safety land." How extensive is "the area" referred to? We know from 2746-2 (p. 196) that Norfolk, Virginia, and "the Tidewater section" is included. But how extensive is the Tidewater section? I don't know, but for the sake of communicating a reasonable guess I've drawn a boundary that separates what I think *might* be safer areas from less safe ones. Thus all boundaries and geophysical data on the map are based (1) on my interpretation of the readings and (2) my interpretation of relevant geological literature. Geophysical information varies greatly from place to place, as does my ability to find and evaluate it.

In working with my "guesstimates" as to general map boundaries and designated potential future geophysical events, readers may wish to obtain more specific and detailed information on geo-hazards from relevant local, state, and federal sources. Also, books like Peter Yanev's *Peace of Mind in Earthquake Country* (Chronicle Books, 1990) may be helpful for those whose property is in an area that may be affected by earthquakes, whether or not the predicted Earth changes occur.

And what do the readings say about just how we may counteract such serious happenings as Earth changes?

Make known the trouble—WHERE IT LIES; that THEY WHO HAVE FORGOTTEN GOD MUST RIGHT ABOUT FACE!

3976-26
April 28, 1941

Tendencies in the hearts and souls of men are such that these [upheavals] may be brought about. For, as indicated through these channels oft, it is not the world, the Earth, the environs about it nor the planetary influences, not the associations or activities, that *rule* man. *Rather* does man—by *his compliance* with divine law—bring *order* out of chaos; or, by his *disregard* of the associations and laws of divine influence, bring chaos and *destructive* forces into his experience.

For *He* hath given, "Though the heavens and the Earth pass away, my *word* shall *not* pass away!" This is often considered as just a beautiful saying, or something to awe those who have been stirred by some experience. But applying them into the conditions that exist in the affairs of the world and the universe in the present, what *holds* them— what are the foundations of the Earth? The word of the Lord!

416-7
October 7, 1935

As pointed out in the main text, the future is not fixed, but conditional. Thus, various features of the looming Earth changes—such as their timing and severity—may be influenced by the positive or negative thoughts and actions of individuals living in the affected areas. *If,* however, many or all of the predictions come to pass, how can we possibly stand up to their adverse consequences? The following readings seem to anticipate our question.

> **Don't think that there will not be trouble, but those who put their trust wholly in the Lord will not come up missing but will find conditions, circumstances, activities, someway and somehow much to be thankful for.**
>
>

1467-18
April 10, 1944

> **These experiences then that have shattered hopes, that have brought disappointments, that have produced periods when there seemed little or nothing in the material life—if they are used ... as stepping-stones and not as those things that would bring resentments, accusations to others, those influences that create discontent, we will find they will become as helpful experiences that may guide the entity ... into a haven that is quiet and peaceful.**
>
> **Hence the necessity ... that the faith ... in a divinity that is WITHIN be held—that shapes the destinies of individual experiences in such a way that the opportunities that come ... are those things which if taken correctly make for the greater soul development.**
>
>

1300-1
November 28, 1936

> **Make thy will then one with His. Be not afraid ... That the periods from the material angle as vi-**

sioned are to come to pass matters not to the soul, but do thy duty TODAY! TOMORROW will care for itself . . . For how hath He given? "The righteous shall inherit the Earth."

Hast thou, my brethren, a heritage in the Earth?

294-185
June 30, 1936

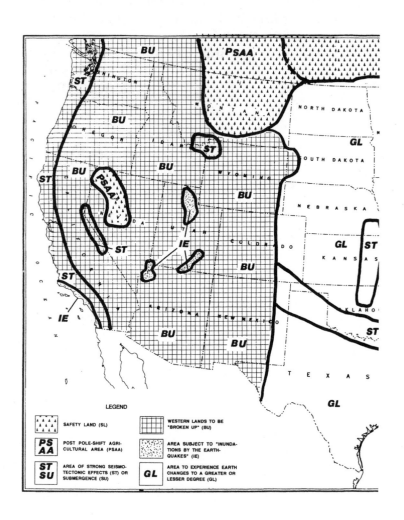

LEGEND

SAFETY LAND (SL)	WESTERN LANDS TO BE "BROKEN UP" (BU)
PS AA — POST POLE-SHIFT AGRI- CULTURAL AREA (PSAA)	AREA SUBJECT TO "INUNDA- TIONS BY THE EARTH- QUAKES" (IE)
ST SU — AREA OF STRONG SEISMO- TECTONIC EFFECTS (ST) OR SUBMERGENCE (SU)	GL — AREA TO EXPERIENCE EARTH CHANGES TO A GREATER OR LESSER DEGREE (GL)

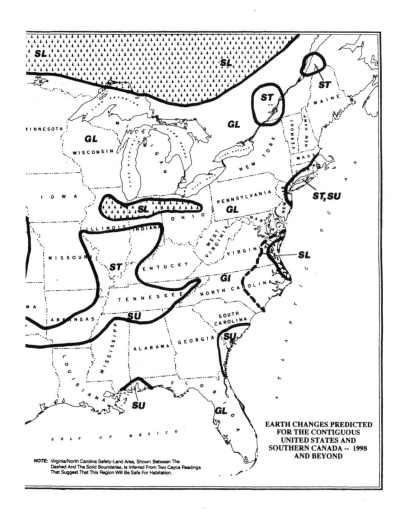

EARTH CHANGES PREDICTED
FOR THE CONTIGUOUS
UNITED STATES AND
SOUTHERN CANADA -- 1998
AND BEYOND

NOTE: Virginia/North Carolina Safety-Land Area, Shown Between The
Dashed And The Solid Boundaries, Is Inferred From Two Cayce Readings
That Suggest That This Region Will Be Safe For Habitation.

Appendix

Deciphering Cayce

Language Used in the Edgar Cayce Readings,
Sources of the Information That Came Through Cayce,
and Errors in the Information

Appendix

Deciphering Cayce

Readers unaccustomed to the unusual language found in the Edgar Cayce readings may find the following explanation[1] by author Gina Cerminara helpful:

> Anyone who has ever had a reading from Edgar Cayce knows that the language in which it was couched was often very hard to understand. A person coming upon the involved phrases and expressions for the first time might well wonder if it is not a case of making the waters muddy so that they may appear to be deep. Such a suspicion is not unnatural; but any person who has made a careful, impartial, and thorough study of the material

knows that it is quite unfounded. Through what one critic called the "rambling, redundant, ambiguous, and evasive verbal meanderings" there ran a current of high and genuine import. Thousands of people whose physical health has been transformed and whose psychological difficulties resolved by the contents of the readings will testify to this fact.

But the question remains: Why WAS the language so odd? Why the psychic double-talk? Why not come out and say, "This is a spade" instead of pussy-footing around with, "This, as we find, has to do with not the consciousness in spirituality (as commonly conceived) but rather the consciousness in materiality, as condensed in what is known as, or called, in the present, an implement of spading, or a spade."

To this question we cannot give a complete authentic answer because we are not in possession of all the facts in the case. But we can draw some reasonable deductions from what the readings themselves have offered as explanation, and we can make an intelligent appraisal of all the facts as we know them, in connection with the phenomenon of language in general.

In the first place, it must be recognized that Mr. Cayce was NOT a medium in the usual sense of the word. It was his own superconscious mind—highly developed and trained in past incarnations in Egypt and Persia—which, when the ordinary consciousness was laid aside in sleep, became active and was able to attune itself with whatever source was necessary to secure the information desired.

The sources upon which the soul-entity of Mr. Cayce could draw were several. These—as explained by the readings themselves—can be summed up as follows: (1) the subconscious mind of Edgar Cayce himself; (2) the subconscious mind of other individuals in the Earth plane (when a life reading was given, the information came in part from the subconscious of the one on whom the reading was given; this was also true, to some extent, in the case of the physical readings); (3) the subconscious minds of discarnate entities in the spirit plane; (4) the soul minds of higher masters; (5) the akashic records; (6) The Universal cosmic consciousness.

Now it seems fairly clear that—whichever source he used to obtain his information, he was speaking from a point of view infinitely vaster and more intricate than the point of view of Earth-plane man, and it is perhaps for this reason that a language difficulty arises. The choice of words is often labored; the phrases seem chosen from an odd and unusual angle—much like a photographic study taken from an angle and in a focus not ordinarily thought of. Exactly what the language mechanism was has never been explained by the readings themselves, except for very fragmentary remarks to the effect that higher-dimensional realities cannot easily be expressed in three-dimensional terms. This sounds reasonable. And piecing together what the readings themselves have said with the linguistic impressions one gathers on examining the readings, one arrives at some fairly satisfactory conclusions.

These impressions can be summarized as follows: (1) the readings sound like a man speaking in a foreign language; (2) they sound like a person from an ancient era trying to speak to a modern era; (3) they sound like a highly educated person—

or an academician—trying to make himself clear to the uneducated—or the nonacademic.

The first impression—that of a man trying to speak in a foreign language—is very strong. One feels, primarily, as if a highly cultivated and benevolent philosopher of a very remote country, say China or Afghanistan, were attempting to speak in English—and the English language is notoriously difficult for foreigners. Native turns of speech, awkward use of idiom, flagrant mistakes of syntax and grammar, oddly combined with occasional perfection of syntax and grammar are typical of the speech of foreigners—who often use subjunctives and pronouns more accurately than we do ourselves, and consequently sound stilted in doing so.

Moreover the use of archaic expressions in the readings, such as "oft," "babe," "wilt," etc., suggest that the speaker is a foreigner not only in space but also in time. These outmoded forms of English may hark back to Mr. Cayce's early American incarnation; and characteristic ways of using certain words, such as "same" and "self" (as, for example, "the entity, remembering same, will make much progress," and "First analyze self") may be idiomatic in one of the ancient languages, either Egyptian or Persian, with which Mr. Cayce's soul-entity was once familiar.

The stiltedness of many passages in the readings gives rise to still a third impression—namely, that of an individual educated far above the level of the person to whom he is talking and attempting to talk down to the level of the second person. This offers both psychological and linguistic difficulties. Ph.D.s and professors, for example (to say nothing of the writers of income tax laws and the Congressional Record) are notorious for their use of style incomprehensible to the majority. In some cases this may be due to affectation or to that kind of social ineptitude that arises from many years of academic seclusion.

But generally speaking, there is among this type of persons a genuine habituation to certain language constructions, which they can depart from only with great effort. The breadth, subtlety, and complexity of their thought is something to which the uneducated are not accustomed. The necessity, then, to make themselves intelligible to people whose outlook is innocent of ideological or technical preoccupations and whose speech is rough and ready for the simple purposes of making a living is often a difficult one.

The difficulty is self-expression, which was obviously experienced by the giver of information in the readings, seems quite comparable. They read like the speech of a man who because of the habitual grandeur of his thoughts, is clumsy when trying to speak at any level other than his own. And, aware of his own clumsiness, he becomes repetitious in the anxious effort to become clear.

Finally, it must be remembered that the readings were given from a point of view of an enlarged consciousness—that is to say, the source of information was conversant with many dimensions and needed to condense what he knew into 3-dimensional terms. This is a difficult thing for us to grasp, because we are all so completely embedded in a 3-dimensional consciousness that we cannot conceive of realities of 4, 5, 6, 7, and more dimensions.

Yet both mathematicians and physicists, occultists and clairvoyants, assure us that higher dimensions DO exist; and the readings themselves, as before indicated, often make reference to the fact.

By way of learning to appreciate the difficulty involved in compressing knowledge into narrower terms than those really adequate to describe it, one might try a little experiment. One might attempt to make a description of something, say the American flag, without using any words that contain the letter "r," on the assumption that you were talking to a person whose version of the English language contained no letter "r."

It seems a simple enough sort of assignment. Yet the moment you began to refer to the most basic features of the thing, you would realize that you could use neither the words STARS nor STRIPES. Your first task would be to find other ways of expressing those ideas. Not even BARS could substitute for STRIPES, and perhaps BANDS would be the closest available equivalent. For STARS you might begin to say, "five-pointed figures," and in the midst of it realize that FIGURES would be incomprehensible to an "r"-less people; so you would settle on "five-pointed symbols of heavenly bodies." RED would be your next preoccupation. You could neither say DARKER nor DEEPER shade of pink, but finally you would approximate the idea by saying "deep pink" or possibly (especially if you were familiar with the nail-polish ads of the country) "passion pink."

Your difficulties are not, however, now at an end. You cannot say there are THIRTEEN bands which ALTERNATE in COLOR, nor can you remark that there are FORTY-eight stars which represent FORTY-eight states of the United States of AMERICA [originally written in 1945]. Even COUNTRY would be unavailable to you.

Devising circumlocutions for all these taboo—or, in a sense, nonexistent words—you might finally emerge with a description something like this: "The flag of the United States of the Continent of the West (that is, the land found by Columbus) consists of eight and five bands of white and a tint that might be called deep pink, passion pink, wine pink, that is to say, the tint of blood—a band of white and then a band of this tint. In the top left of the flag is a field of blue on which we see twenty and twenty-eight five-pointed symbols of heavenly bodies, each to stand as symbol of the twenty and twenty-eight states that make up the land."

This—as anyone who has ever read a Cayce reading will admit—sounds extraordinarily like a reading—and its clumsiness very likely arises from the same basic cause.

The readings, of course, need no apology. Their merit has been proven so many thousands of times in so many extraordinary cases that the genuineness of their clairvoyance can simply not be questioned. But it has seemed worthwhile to write this article in the interest of bringing out into the open a matter which, to some people at least, looms large as an obstacle to the full acceptance of the readings.

For general publication purposes, indeed, it has become necessary to adopt careful editorial policies of simplification or clarification, in order that the message may be understandable by the majority; but even after this has been done, under rigorous supervision of Mr. Cayce's secretary, the language still retains a certain quaint flavor. In this connection it is well to remember, however, that the substance, sincerity, or intrinsic value of a message cannot always be judged by the language in which it is clothed. Some of the best writing in the civilized world, for example, is now being done in the field of advertising. The most poetic of prose—sensitive, direct, dynamic—is found in advertising copy, and frequently the pages of advertisements make more compact, interesting, and instructive reading than the text of the magazine itself. But oftener than not that magnificent writing is the cloak for a lying message, a hypocritical appeal, and a subtle, deliberate, and crassly self-interested playing upon the reader's sensibilities.

By all means use all your faculties of discrimination and critical judgment with regard to the Cayce readings. But do not permit the clumsiness of the language to deflect you from the genuine worth of their contents.

As for errors in the information obtained by Cayce, an author and student of the readings, W. H. Church, made the following observations,[2] as based upon a 1933 reading on this very subject. Among the causes for errors or lapses, one was the unwillingness of the body-consciousness of the recipient of the reading, or of Cayce himself, to be properly responsive at the time. Ill health was another cause, as were the mental attitudes of those in Cayce's presence at the time the information was being sought. "Combative influences" in the experience of the entity seeking or "the continual warring" between flesh and spirit were also contributors to errors or lapses in communication. The reading also said that the akashic records of a soul entity are either positive or negative—positive being "good" and negative being "error."

Finally, in 1971, the two sons of Edgar Cayce, Edgar Evans and Hugh Lynn, respectively, published a book about their father entitled, *The Outer Limits of Edgar Cayce's Power*,[3] in which they disclosed a number of reasons for errors and lapses in the readings. They pointed out that an obvious criterion for the validity of any psychic information is its source. And on different occasions Cayce seemed to obtain information from a variety of sources. In a chapter entitled, "Was Edgar Cayce's Story a Hoax?" they tried to quantify the reliability of Cayce's readings by analyzing reports for a sample of 150 randomly selected readings out of the 14,306 readings on file. In this random group, covering readings from 1910 to 1944, there were 110 physical readings, 34 life readings, 4 business readings, 1 dream reading, and 1 aura-chart reading. No reports were obtained from 74 of the reading recipients. Eleven of the reports were negative; 65 reports were favorable. It was found that Cayce had an "accuracy" of 85 percent, for the types of readings described. Because most of the readings considered in their study were physical readings, Cayce's sons concluded that this level of accuracy "compares favorably to that of modern physicians" (p. 24; see footnote 3).

Notes

Introduction
1. Kerr, R.A., "Briefing: Ancient River System Across Africa Proposed," *Science*, v. 233, p. 940.
2. These CD-ROMs may be purchased from the A.R.E. Press (1-800-723-1112).

Chapter One
Cayce's Psychic Visions
1. U.S. Weather Bureau, *Monthly Weather Review*, v. 54, no. 10 (October 1926), pp. 409-452.
2. India Meteorological Department, Calcutta: *India Weather Review*, 1926, 311 pp.
3. Earthquake Data Base System, U.S. Geological Survey, National Earthquake Information Center, Golden, Colorado.
4. U.S. Coast and Geodetic Survey, "Earthquake History of California and Western Nevada," Serial No. 609, Washington, 1951, 35 pp.
5. See Figure 27 of the California Division of Mines and Geology's Open File Report 82-17 SAC, *Areas Damaged by California Earthquakes, 1900-1949,* published in 1982.
6. *New York Times*, August 8, 1995, "Hopes for Predicting Earthquakes, Once So Bright, Grow Dim," p. B8.
7. Heck, N.H., "List of Seismic Sea Waves," *Bull. Seismological Soc. Amer.,* Oct. 1947, no. 4, pp. 269-284.
8. The fact that Search for God Study Group No. 1 rejected Halaliel's offer to direct the materials given to that group in a clearer and more organized fashion does *not* require us to reject the information provided in 3976-15. Such has been implied by geologist John Peterson, as cited in John White's book on pole shift (see White, J., 1994, *Pole Shift,* A.R.E. Press, Virginia Beach, Va., p. 405). We, however, can come to our own conclusions, as based upon the readings quoted in the text.
9. See p. 25 of *Earth Changes Update,* by H.L. Cayce. As counterpoint to Hugh Lynn's position that a change in attitudes in Alabama mitigated the predictions of sinking in the extreme

southwestern and northwestern parts of the state, we must remember the negative attitudes in Alabama that led to George Wallace's election as governor in 1963. And consider the article in the *New York Times* for September 28, 1980, entitled "Klan Group in Alabama Training for 'Race War.' " It mentions a Decatur, Ala., gun battle in which a black man shot a former Exalted Cyclops of the Klan and three other people were wounded. Decatur is in northwestern Alabama, where "citizens militias" are active today. More recently, the *New York Times*'s Ronald Smothers wrote (January 20, 1996) about a Federal investigation into the burning of rural black Baptist churches in west-central Alabama, and *Time* magazine's Jack E. White wrote (3/18/96) about additional church torchings in Alabama designed to intimidate black people, including a black judge.

10. See p. 36 of *Visions and Prophecies for a New Age*, by M. A. Thurston, 1981, A.R.E. Press, Virginia Beach, Va.

11. The biblical references are " . . . where she [Earth] is nourished for a time, and times, and half a time . . . " (Rev. 12:14); " . . . until a time and times and the dividing of time" (Dan. 7:25); and " . . . it *shall be* for a time, times, and an half . . . all these *things* shall be finished." (Dan. 12:7)

Chapter 2
Pole Shift

1. Loper, D.E., et al., 1988, "A Model of Correlated Episodicity in Magnetic-Field Reversals, Climate, and Mass Extinctions," *Jour. Geology*, v. 96, no. 1, pp. 1-11.

2. As quoted in DePaolo, et al., 1991, "Physics and Chemistry of Mantle Plumes," *EOS*, May 21, p. 236; copyright by the American Geophysical Union.

3. Modified from Fig. 1 of DePaolo, et al., in footnote immediately above.

4. As described in Runcorn, 1960, *Methods and Techniques in Geophysics*, Interscience Pub. Inc., New York, 374 pp.

5. Courtillot, V., and Besse, J., 1987, "Magnetic Field Reversals, Polar Wander, and Core-Mantle Coupling," *Science*, v. 237, pp. 1140-1147.

6. See Lewin, R., 1983, "What Killed the Giant Mammals?" *Science,* v. 221, p. 1036. (Others think that Clovis Man, the big-game hunting culture that flourished around 10,900 to 11,300 years ago, killed the animals, as discussed in Lewin's report.)

7. See *Nature,* 1995, v. 387, p. 23.

8. See Bonatti, E., 1994, "The Earth's Mantle Below the Oceans," *Scientific American,* pp. 44-51.

9. Harrison, W., 1958, "Marginal Zones of Vanished Glaciers," *Jour. Geology,* v. 66, pp. 72-95.

10. Rampino, M.R., et al., 1979, "Can Rapid Climate Change Cause Volcanic Eruptions?" *Science,* v. 206, pp. 826-828.

11. James, T.S., and Ivins, E.R., 1995, "Present-Day Antarctic Ice Mass Changes and Crustal Motion," *Geophysical Res. Letters,* v. 22, pp. 973-976.

12. Blakenship, D.D., et al., "Active Volcanism Beneath the West Antarctic Ice Sheet," *Nature,* v. 361, 11 Feb., p. 526.

13. Williams, S.N., 1995, "Erupting Neighbors—at Last," *Science,* v. 267, January 20.

14. Locke, W.W., and Meyer, G.A., 1994, "A 12,000-Year Record of Vertical Deformation Across the Yellowstone Caldera Margin," *Jour. Geophysical Res.,* v. 99, no. B10, Oct. 10, p. 20,079.

Chapter 3
Have the Earth Changes Begun?

1. Westaway, R., 1993, "Quaternary Uplift of Southern Italy," *Jour. Geophys. Res.,* v. 98, no. B12, pp. 21,741-21,772.

2. Bevis, M., et al., 1995, "Geodetic Observations of Very Rapid Convergence and Back-Arc Extension at the Tonga Arc," *Nature,* v. 374, pp. 249-251.

3. Giardini, D., and Lundgren, P., 1995, "The June 9 Bolivia and March 9 Fiji Deep Earthquake of 1994: II Geodynamic Implications," *Geophys. Res. Letters,* v. 22, no. 16, pp. 2281-2284.

4. Anonymous, 1994, "An Island Dies, an Empire Falls," *Discover,* April, p. 14.

5. Prevot, R., et al., 1994, "A Shallow Double Seismic Zone Beneath the Central New Hebrides (Vanuatu)," *Geophys. Res. Letters,* v. 21, no. 19, pp. 2159-2162, and Taylor, et al., 1995, "Geodetic Measurements of Convergence at the New Hebrides

Island Arc Indicate Arc Fragmentation Caused by an Impinging Aseismic Ridge," *Geology,* v. 23, no. 11, pp. 1011-1014.

6. Broecker, W.S., 1995, "Chaotic Climate," *Scientific American,* November, pp. 62-68.

7. Parrilla, G., et al., 1994, "Rising Temperature in the Subtropical North Atlantic Ocean over the Past 35 Years," *Nature,* v. 369, pp. 48-50.

8. Young, S., 1994, "Insects That Carry a Global Warning," *New Scientist,* April 30, pp. 32-34.

9. Serreze, M.C., et al., "Diagnosis of the Record Minimum in Arctic Sea Ice During 1990," *Geophys. Res. Letters,* v. 22, no. 16, pp. 2183-2186.

10. Anonymous, "Polar Ice on the Wane," *Science News,* v. 148, August 19, 1995, p. 123.

Chapter 4
Is America Vulnerable?

1. Press, F., and Allen, C., 1995, "Patterns of Seismic Release in Southern California," *Jour. Geophys. Res.,* v. 100, no. B4, pp. 6421-6430.

2. Scientists of the USGS and S. Cal. Earthquake Center, 1994, "The Magnitude 6.7 Northridge, California, Earthquake of 17 January 1994," *Science,* v. 266, pp. 389-396.

3. Fumal, T.E., et al., 1993, "A 100-Year Average Recurrence Interval for the San Andreas Fault at Wrightwood, California," *Science,* v. 259, pp. 199-203.

4. Kerr, R.A., 1995, "Bigger Jolts Are on the Way for Southern California," *Science,* v. 267, pp. 176-177.

5. Hatch, M.E., and Schug, D.L., 1995, "Evidence for Active Faulting in Downtown San Diego," *Abstr. Vol.,* Annual Meet., Assoc. Engineering Geol., Sacramento, Oct. 2-8, p. 54.

6. Kerr, R.A., 1994, "How Many More After Northridge?" *Science,* v. 263, p. 460.

7. Kerr, R.A., 1989, "Reading the Future in Loma Prieta," *Science,* v. 246, p. 436.

8. See pp. 187 and 189. *Assembling California* was published by Farrar, Straus, and Giroux, N.Y., in 1993.

9. Crowell, J.C., 1987, "The Tectonically Active Margin of the

Western U.S.A.," *Episodes,* v. 10, no. 4, pp. 278-282.

10. DeMets, C., 1995, "A Reappraisal of Sea-Floor Spreading Lineations in the Gulf of California," *Geophys. Res. Letters,* v. 22, no. 24, pp. 3545 -3548.

11. Prims, J., and Furlong, K., 1995, "Subsidence of San Francisco Bay: Blame It on Salina," *Geology,* v. 23, no. 6, pp. 559-562.

12. U.S. Geol. Survey, 1992, "Special Issue: The Cape Mendocino Earthquake of April 25-26, 1992," *Earthquakes and Volcanoes,* v. 23, no. 3, p. 92.

13. Wuethrich, B., 1994, "It's Official: Quake Danger in Northwest Rivals California's," *Science,* v. 265, p. 1802.

14. Heatong, T.H., and S.H. Hartzell, 1987, "Earthquake Hazards on the Cascadia Subduction Zone," *Science,* v. 236, pp. 162-168.

15. *Science News,* April 8, 1995, "Waves in the Night: Clues to a Quake," p. 223.

16. Atwater, B.F., and Moore, A.L., "A Tsunami About 1000 Years Ago in Puget Sound, Washington," *Science,* v. 258, pp. 1614-1617.

17. Ryall, A., et al., 1966, "Seismicity, Tectonism, and Surface Faulting in the Western United States During Historic Time," *Bull. Seismological Soc. Amer.,* v. 56, pp. 1105-1135.

18. U.S. Geol. Survey, 1988, "Assessing Earthquake Hazards of Urban Areas," *Earthquakes and Volcanoes,* v. 20, no. 6, pp. 209-210.

19. Center for Earthquake Studies, 1992, *Damages and Losses from Future New Madrid Earthquakes.* Southeast Missouri State University, Cape Giradeu, MO 63701, 34 pp., 4 append. (The author of the report is David M. Stewart, Ph.D., Route 1, Box 646, Marble Hill, MO 63764.)

20. "Evidence Found of Prehistoric Quakes," 1992, *Geotimes,* October, p. 6.

21. Allen, W., 1994, "FEMA Study Predicts Casualties, Destruction by Area Earthquake," *St. Louis Post-Dispatch,* Sunday, Jan. 23.

22. Kerr., R.A., 1985, "Signs of an Eastern Quake?" *Science,* Nov. 1, p. 531.

23. Newman, W.S., 1987, "Holocene Neotectonics and the Ramapo Fault Zone Sea-Level Anomaly," *Sea-Level Fluctuation and Coastal Evolution*, Soc. Econ. Paleontol. and Mineral., Spec Publ. 41, Tulsa, pp. 106-110.

24. Hutchinson, Dr. R., and Grow, J.A., 1985, "New York Bight Fault," *Geol. Soc. Amer. Bull.*, v. 196, pp. 975-989.

25. Brown, L.D., 1978, "Recent Vertical Crustal Movement Along the East Coast of the United States," *Tectonophysics*, v. 44, pp. 205-231.

26. Anderson, W.A., et al., "Crustal Warping in Coastal Maine," *Geology*, v. 12, pp. 677-680.

27. Barosh, P.J., 1986, "Neotectonic Movement, Earthquake and Stress State in the Eastern United States," *Tectonophysics*, v. 132, pp. 117-152. (Fig. 16 of this book is adapted from Fig. 10 of the P.J. Barosh paper, with kind permission of Elsevier Science-NL, Sara Burgerhartstraat 25, 1055 KV Amsterdam, The Netherlands.)

28. See footnote 23, p. 131, and Fig. 16.

29. Phylos, 1952, *A Dweller on Two Planets*, Borden Pub. Co., Los Angeles. (Phylos's forecast for the United States is on pp. 435-439 of the 1952 edition.)

30. *Chicago Tribune*, series of three articles, November 6-8, 1994.

31. *Wall Street Journal*, Oct. 25, 1994. "U.S. Hits Haiti with Double Whammy."

32. *Wall Street Journal*, July 11, 1994.

33. *Wall Street Journal*, July 13, 1994.

34. Klare, M.T., 1984, *American Arms Supermarket*, Univ. Texas Press, Austin. (See Table 1.)

35. NBC Evening News, Aug. 30-31, 1994. Also, a recent Pentagon study predicts that the U.S. will probably hold its position as the world's biggest arms supplier for the foreseeable future (*Wall Street Journal*, February 24, 1995, p. A7).

36. Kaplan, R.D., 1994, "The Coming Anarchy," *Atlantic Monthly*, February, pp. 44 ff.

37. *New York Times*, Dec. 15, 1995, "Extremist Army Group Wages War with U.S. Policy."

38. Posner, Gerald, 1993, *Case Closed: Lee Harvey Oswald and*

the Assassination of JFK, Random House, New York.

39. U.S. Bureau of the Census, *Statistical Abstract of the United States, 1993,* "Work Stoppages: 1960 to 1992," Table 686.

40. Epstein, G., 1996, "Despite Years of Stagnation for Labor, Will Wage Inflation Revive?" *Barron's,* February 5, p. 54.

41. See G. Malloan, "The Anatomy of Breakdowns in the Social Order," *New York Times,* January 13, 1998.

42. Fellowship of the Inner Light, 1977, *Earth Changes and the New Planet Earth,* The Masters Press, P.O. Box 206, Virginia Beach, VA 23458.

43. Goodman, J., 1978, *We Are the Earthquake Generation: Where and When the Catastrophes Will Strike,* Seaview Books, New York.

44. Auster, L., 1993, *The Path to National Suicide: An Essay on Immigration and Multiculturalism.* See also, Jordan, Barbara, "The Americanization Ideal," *New York Times,* Sept. 11, 1995.

45. Buchanan, P.J., 1992, "Yes, Mario, There Is a Culture War," *Chicago Tribune,* Sept. 14, 1992.

46. Davidson, J.D., 1994, "The New Form of Politics: Drug Lords and Waning of the Nation State," *Strategic Investment,* Oct. 26, Agora, Inc., 824 Baltimore St., Baltimore, MD 21202. (For more on insights by James Dale Davidson, call 800-433-1528 and ask about his monthly newsletter, *Strategic Investment.*)

47. Davidson, J.D., July 25, 1995, "This Month: 'Conquest of the U.S. by Mexico,' " *Strategic Investment.* See footnote 46.

48. *Wall Street Journal,* July 10, 1995, p. A12.

49. *Chicago Tribune,* September 24, 1995.

50. Lead editorial in *Wall Street Journal,* December 5, 1994.

51. *Chicago Tribune,* August 20, 1995.

52. *Wall Street Journal,* December 22, 1994.

53. *Chicago Tribune,* September 24, 1995.

54. R.I. Friedman, "The Money Plane," *New York Magazine,* January 22, 1996.

55. Williams, P., and Woessner, P.N., 1996, "The Real Threat of Nuclear Smuggling," *Sci. American,* January, pp. 40-44.

56. As suggested, for example, in *The Men Who Killed Kennedy*

(Nov. 1995), a six-hour documentary on Direct TV Corp.'s History Channel, hosted by Roger Mudd. Mafia mobsters are said to have had Kennedy killed because they were losing out on gambling profits in Cuba and elsewhere, due to U.S. government anti-Mafia actions in the Kennedy administration. This documentary and its one-hour sequel, subtitled *The Truth Shall Make You Free,* has been thoroughly criticized by Dorothy Rabinowitz in the *New York Times* (4/1/96). She points out that the sequel does not even mention the Mafia conspiracy that formed such a prominent part of the original "conspiracy gala." The controversy continues over a possible mob connection to the Kennedy assassination.

Chapter 5
Will Japan Be Submerged?

1. Tanioka, Y., et al., 1993, "Unusual Rupture Process of the Japan Sea Earthquake," *EOS,* v. 74, August 24; copyright by the American Geophysical Union.

2. Tanioka, Y., et al., 1995, "The Great Kurile Earthquake of October 4, 1994, Tore the Slab," *Geophys. Res. Letters,* July 1, pp. 1661-1664. This most recent major earthquake in the ocean to the east of Japan occurred along the southernmost end of the Kurile trench and registered M 7.8. The quake and related tsunami caused great damage to Shikotan Island. It was due to the subducting Pacific plate underthrusting northernmost Japan and the southern Kurile islands.

3. Anon., 1978, "Japan Trench Transected," *Geotimes,* April, pp. 16-20.

4. *Science,* 4 December 1992, p. 1576.

5. 1994, ABC-TV, *Earthquakes: The Terrible Truth,* program aired May 21.

6. From an article by the *Boston Globe's* Colin Nickerson, entitled "Killer Quake: Tokyo's Next Big One," that appeared in the *Chicago Tribune* for Sunday, November 20, 1988, Section 5, p. 13.

Chapter 6
Shaky Latin America and the Southern Hemisphere
1. Kerr, R.A., 1994, "Bolivian Quake Deepens a Mystery," *Science*, v. 264, p. 1659.

Chapter 7
Lost Continents
1. A transform fault is a special variety of strike slip fault in which movement is parallel to the fault's strike. But along a transform fault, the displacement suddenly stops or changes form. Ocean floors are full of transform faults, mostly in association with mid-ocean ridges.

2. McCunn, H.J., 1973, "Vertical Uplift Explanation for Plate Tectonics," *Amer. Assoc. Petrol. Geol. Bull.*, v. 57, pp. 1644-1657.

3. Zielinski, G.A., et al., 1996, "An 110,000-Year Record of Explosive Volcanism from the GISP2 (Greenland) Ice Core," *Quaternary Research*, v. 45, no. 2, pp. 109-118.

4. Gronvold, K., et al., 1995, "Ash Layers from Iceland in the Greenland GRIP Ice Core Correlated with Oceanic and Land Sediments," *Earth and Planet. Sci. Letters*, v. 135, pp. 149-155.

5. See *The Geology of North America*, Vol. M, 1986, "The Western North Atlantic Region," edited by P.R. Vogt and B.E. Tucholke, Geol. Soc. America, Boulder, Colorado.

6. Shepard, F.P., 1963, *Submarine Geology*, Harper and Row, New York.

7. See *The Geology of North America*, Vol. I-2, 1988, "The Atlantic Continental Margin: U.S.," edited by R.E. Sheridan and J.A. Grow, Geol. Soc. America, Boulder, Colorado.

8. Sheridan, R.E., et al., 1981, "Stratigraphy and Structure of the Southern Blake Plateau, Northern Florida Straits, and Northern Bahamian Platform," *Amer. Assoc. Petrol. Geol. Bull.*, v. 65, no. 12, pp. 2571-2593.

9. Ferro, R., and Grumley, M., 1970, *Atlantis*, Doubleday, New York.

10. Harrison, W., 1971, "Atlantis Undiscovered—Bimini, Bahamas," *Nature*, v. 230, pp. 287-289.

11. McKusick, M., and Shinn, E. A., 1980, "Bahamian Atlantis Reconsidered," *Nature*, v. 287, pp. 11-12.

12. ODP Leg 153 Shipboard Scientific Party, 1995, "Probing

the Foundation of the Mid-Atlantic Ridge," *EOS,* v. 76, no. 13, March 28, p. 129.

13. The Ministry of Lands and Works in Nassau will want to have full details of the project, including plans, calculations, names and addresses of interested parties, and a general description of the project and its objectives before granting permission for exploratory dredging for prehistoric artifacts. A typical dredging specification might include making a nominal mile-long cut, 50 feet wide, with a 10-inch dredge, to a 30-foot depth. The spoil would be pumped to shore for screening on the beach. (This would provide useful nourishment of the beach.)

14. Johnson, A.C., and Swain, C.J., 1995, "Further Evidence of Fracture-Zone Induced Tectonic Segmentation of the Antarctic Peninsula from Detailed Aeromagnetic Anomalies," *Geophys. Res. Letters,* v. 22, pp. 1917-1920.

Chapter 8
Danger Lands/Safety Lands

1. Thurston, M.A., 1981, *Visions and Prophecies for a New Age,* A.R.E. Press, Virginia Beach, VA 23451, 228 pp.

Chapter 9
Humanity's Response to Earth Changes

1. Boyd, D., 1974, *Rolling Thunder,* Random House, New York. Also, Boyd says (p. 50), "I have since come to know some part of Indian prophecy that is preserved chiefly by the Hopi people and maintained by traditional-Indian spokesmen everywhere. The currently significant part of those prophecies pertains to an approaching transition that is often called 'the day of purification.' This prophecy coincides with the claims of ecologists and scientists who believe that imbalance in nature has passed the point of no return." (Note that this was written before the catastrophe at Chernobyl and the torching of Kuwait's oil fields. Thus, the world's ecosystems become even more insulted as time passes.)

2. Collin, R., 1954, *The Theory of Celestial Influence: Man, the Universe, and Cosmic M,ystery,* Vincent Stuart, London, 393 pp.

3. Advisory Comm. for Internat. Decade for Natural Hazard Reduction, 1987, *Confronting Natural Disasters,* Nat. Acad. Press, Washington, D. C.

4. Schwartz, S.I., 1995, "Four Trillion Dollars and Counting," *Bull. Atomic Sci.,* Nov./Dec., pp. 32-51. Reprinted by permission of the Bulletin of the Atomic Scientists, copyright © 1995 by the Educational Foundation for Nuclear Science, 6042 South Kimbark Avenue, Chicago, IL 60637, USA. A one-year subscription is $36.00.

5. Codevilla, A.M., 1995, "How the Missile Shield Was Sabotaged," *New York Times,* Mar. 4.

6. Park, R.L., 1995, "Stars Wars: The Sequel," *New York Times,* Feb. 10.

7. Lienkaemper, J.J., et al., 1991, "Historic Creep Rate and Potential for Seismic Slip Along the Hayward Fault, California," *Jour. Geophys. Res.,* v. 96, pp. 18,261-18,283.

8. Leventhal, P.L., 1994, "The New Nuclear Threat," *Wall Street Journal,* June 8.

9. See p. 6 of the Japanese Ministry of Foreign Affairs' promotional piece entitled, "Striving for Long-Term Energy Security: Japan's Policy on the Use of Plutonium," Feb. 1995. See also D.P. Hamilton's "Japan's Plutonium Program Won't Die," *Wall Street Journal* (4/3/96), in which Hamilton says, "Plans for a huge plant that will reprocess nuclear waste in northern Japan [at Rokkasho] are still rolling forward" and that Prime Minister Hashimoto "recently called on his cabinet ministers to find new ways to restore 'public trust' in the nation's nuclear program."

10. Hatsumi, R., "What Aum Offered," *New York Times,* OP-ED page, May 24, 1995.

11. Williams, M., 1995, "Japan Attempts to Squelch Article to Protect Official," *Wall Street Jour.,* Nov. 20.

12. *New York Times,* Dec. 15, 1995, "Circling the Wagons, China's Aging Leadership Swaggers in Its Growing Isolation from the West."

13. Leventhal, P., and Horner, D., 1995, "Peaceful Plutonium? No Such Thing," *New York Times,* Jan. 25, OP-ED page.

14. See Dower, J.W., 1993, *Japan in War and Peace,* New Press,

N. Y., Chapter 3, " 'NI' and 'F': Japan's Wartime Atomic Bomb Research," and Dower's quote (p. 56) of D. Shapley's Jan. 13, 1978, *Science* article in which she says that the historical importance of the project (Japan's attempt to build an atom bomb) "lies not in the fact that Japan failed but that she tried, and that Japan's postwar attitude, that she, as the one nation victimized by atomic weapons, is above seeking to acquire them for herself, is not historically accurate . . . [and] . . . if other factors had made a bomb possible, the leadership—which by the end of the war were placing their own youth in torpedoes to home them in on the advancing U.S. Fleet—would not have hesitated to use the bomb against the United States."

15. "Germans Reject Demand to Alter Nuclear Plant," *New York Times,* January 21, 1996.

16. Wald, M.L., 1996, "U.S. to Resume Reprocessing of Nuclear Fuel: Energy Department Says Recovery Is the Safest Plan for Plutonium," *New York Times,* Jan. 5.

17. *New York Times,* Dec. 15, 1995, "U.S. Suspects India Prepares for Nuclear Test."

Chapter 10
A New Age, A New Understanding

1. As stated by Robert Ornstein in his book *The Mind Field* (1996, Malor Books, p. 24), "According to esoteric tradition, the 'organ of perception,' which can be tutored in the same fashion as is language, is what we term intuition. Although the phrase is often maligned, conventionally used to indicate random guesswork or a mysterious combination of elements, it should be properly understood as *knowledge without recourse to inference.*"

2. Shah, Idries, 1994, *The Commanding Self,* Octagon Press, London, 332 pp. See p. 90, "Right and Wrong Study." Shah's definition of the commanding self is taken from the book jacket.

3. Sprott, W.H.J., 1966, *Human Groups*, Penguin Books, Baltimore, Md., 201 pp.

4. Deikman, A.J., 1988, *Evaluating Spiritual and Utopian Groups,* Inst. for Cultural Research, Monog. 25, 16 pp. Avail-

able from the Institute at P. O. Box 13, Tunbridge Wells, Kent, TN3 OJD, England.

5. Available from the A.R.E. Press (1-800-723-1112).

6. *Step-by-Step: A Guide to* A Search for God, Book I, available from the A.R.E. Press (1-800-723-1112). The A.R.E. Study Group Department can be contacted at P.O. Box 595, Virginia Beach, VA 23451-0656. (Ph. 1-757-428-3588.) A helpful earlier book is: Kidd, Worth, 1971, *Edgar Cayce and Group Dynamics,* A.R.E. Press, Virginia Beach, Va., 75 pp.

Addendum

1. See note 8 on page 305. Also, an interesting article on "The Halaliel Question," by W. H. Church, can be found in the May/June, 1992, issue of *Venture Inward.*

2. See, for example, Peterson, John, 1992, "Earth Changes: An Alternative View," *Venture Inward,* May/June issue, p. 36.

3. On the CD-ROM for Cayce's readings we find, under "Reports of Reading 3976-15," item R3:

In re. 3976-15, Par. 5-7, consider this extract from Smith's Bible Dictionary, *A. J.Holman & Co., Ltd., 1895 (a copy of which Edgar Cayce bought in 1896!):*

PEN'IEL, PENUEL (face of El-God). (Gen. xxxii 30).

Where Jacob wrestled with a man who changed Jacob's name to ISRAEL. It does not appear again until after 500 yrs. when Gideon, on his way from Succoth, on the Jordan, chasing Zeba and Azlmunna, being faint from want of food, asked the people of this place for bread for his soldiers, and was denied (Judg. viii. 8). He destroyed the tower of the city on his return (ver. 17). Jeroboam rebuilt the place (1 K. xii. 25). It has never been mentioned since, and is now lost.

4. See *Pole Shift,* by John White, p. 205 (reference no. 8, p. 305, of this book).

5. See p. 37 of John Peterson's article (reference no. 2 above).

6. *The Oxford Dictionary and Thesaurus: American Edition,* 1996, Oxford Univ. Press, N.Y., p. 610.

7. Monastersky, R., 1997, "The Call of Catastrophes," *Science News Anniversary Supplement*, p. S20.

8. Cayce, E. E., and others, 1997, *Mysteries of Atlantis Revisited*, St. Martin's Press, New York, 212 pp.

9. Richards, P. G., and X. Song, "Earth's Inner Core: The Discoveries of Recent Decades," *EOS Trans. AGU*, Spring Meet. Suppl., U31C-13, 1997.

10. Steinberger, B., and R. J. O'Connell, 1997, "Changes of the Earth's Rotation Axis Owing to Advection of Mantle Density Heterogeneities," *Nature*, v. 387 (May 8, p. 169).

11. Van der Hilst, and others, 1997, "Evidence for Deep Mantle Circulation from Global Tomography," *Nature*, v. 386, p. 578.

12. Chang, W. L., and R. B. Smith, 1996, *American Geophysical Union Fall Meeting Proceedings*, Abstract No. S31C-6. A normal-faulting earthquake is one caused by sudden movement of the overlying side of a fault, downward relative to the underlying side. The angle of a normal fault-plane is between 45° and 90°.

13. Lander, J. F., and others, 1993, *Tsunamis Affecting the West Coast of the United States*, National Geophysical Data Center, Boulder, CO 80303, 242 pp.

14. *New York Times*, "After 6 Years of Civil War, a Lawless Liberia," Feb. 1, 1996.

15. "Collapse of the Russian Army," *Strategic Investment*, Apr. 16, 1997.

16. J. Wheeler, in *Strategic Investment*, Apr. 16, 1996.

17. *New York Times*, "Editorial: Reckless Salesmanship on Arms," Apr. 21, 1997.

18. *Washington Post*, in *The Virginian-Pilot*, "U. S. Tries to Swell the Persian Gulf's Bulging Arsenals," Apr., 1997.

19. *New York Times*, "Costs of the C.I.A.," Apr. 25, 1997.

20. El-Fiky, G. S., and others, 1996, "Vertical Crustal Movement in the Tohoku District, Japan, Deduced from Dynamic Adjustment of Leveling and Tidal Data," *Bulletin of Earthquake Research Institute, Univ. Tokyo*, v. 71, pp. 47-71.

21. *New York Times*, "Burning of Amazon Picks Up Pace, with Vast Areas Lost," Sept. 12, 1996.

22. Pearce, F., 1996, "Toxic Legacy of Britain's Oil Boom," *New Scientist*, Dec. 7.

23. "NIF Ignites Changes at Livermore," *Science*, v. 275, p. 1252, Feb. 28, 1997.

24. Amelung, F., and G. King, 1997, "Large-Scale Deformation Inferred from Small Earthquakes," *Nature*, v. 386, p. 702 (April 17).

25. These "hierarchies" seem to be the same as those referred to on p. 229, paragraph no. 5.

26. *Wall Street Journal*, "U. S. Discloses Long-Secret Nuclear Fiascoes," Jan.

27. *The Virginian-Pilot*, "N. Korea Has Nuclear Arms, High-Ranking Defector Says," Apr. 23, 1997.

28. Wheeler, J., 1996, "Behind the Lines," *Strategic Investment*, Oct. 16, p. 7.

29. Tyler, P. E., "China Raises Nuclear Stakes on the Subcontinent," *New York Times*, Aug. 27, 1996.

30. Milhollin, G., "Opinion: China Cheats (What a Surprise!)," *New York Times*, Apr. 24, 1997.

31. Hackett, J. T., "Russia's Big Bang," *Wall Street Journal*, Mar. 28, 1997.

32. Burke, J., and R. Ornstein, 1995, *The Axemaker's Gift*, Grosset/Putnam, N.Y.

33. Malone, T. W., from his statement on the book jacket for the reference cited just above.

34. This paragraph is paraphrased from the advertising for the book.

35. See reference note 2, p. 316. The quote is from p. 213 of Shah's book, 1996.

36. TEXT OF READING 5753-1

This psychic reading was given by Edgar Cayce at his home, on the 16th day of June, 1933, before the Second Annual Congress of the Association for Research and Enlightenment, Inc., in accordance with request by those present.

PRESENT

Edgar Cayce; Gertrude Cayce, Conductor; Gladys Davis, Steno. And approximately thirty-five other people attending the Congress.

READING

1 GC: You will give at this time a comprehensive discourse on reincarnation. If the soul returns to the earth through a succession of appearances, you will explain why this is necessary or desirable and will clarify through explanation the laws governing such returns. You will answer the questions which will be asked on this subject.

2 EC: Yes. In giving even an approach to the subject sought here, it is well that there be given some things that may be accepted as standards from which conclusions—or where parallels—may be drawn, that there may be gathered in the minds of those who would approach same some understanding, some concrete examples, that may be applied in their own individual experience.

3 Each soul that enters, then, must have had an impetus from some beginning that is of the Creative Energy, or of a first cause.

4 What, then, was—or is—the first cause; for if there be law pertaining to the first cause it must be an unchangeable law, and is—IS—as "I AM that I am!" For this is the basis from which one would reason:

5 The first cause was, that the created would be the companion for the Creator; that it, the creature, would—through its manifestations in the activity of that given unto the creature—show itself to be not only worthy of, but companionable to, the Creator.

6 Hence, every form of life that man sees in a material world is an essence or manifestation of the Creator; not the Creator, but a manifestation of a first cause—and in its own sphere, its own consciousness of its activity in that plane or sphere.

7 Hence, as man in this material world passes through, there are the manifestations of the attributes that the consciousness attributes to, or finds coinciding with, that activity which is manifested; hence becomes then as the very principle of the law that would govern an entrance into a manifestation.

8 Then a soul, the offspring of a Creator, entering into a consciousness that became a manifestation in any plane or sphere of activity, given that free-will for its use of those abili-

ties, qualities, conditions in its experience, demonstrates, manifests, shows forth, that it reflects in its activity towards that first cause.

9 Hence in the various spheres that man sees (that are demonstrated, manifested, in and before self) even in a material world, all forces, all activities, are a manifestation. Then, that which would be the companionable, the at-oneness with, the ability to be one with, becomes necessary for the demonstration or manifestation of those attributes in and through all force, all demonstration, in a sphere.

10 Because an atom, a matter, a form, is changed does not mean that the essence, the source or the spirit of it has changed; only in its form of manifestation, and NOT in its relation with the first cause. That man reaches that consciousness in the material plane of being aware of what he does about or with the consciousness of the knowledge, the intelligence, the first cause, makes or produces that which is known as the entering into the first cause, principles, basis, or the essences, that there may be demonstrated in that manifested that which gains for the soul, for the entity, that which would make the soul an acceptable companion to the Creative Force, Creative Influence. See?

11 As to how, where, when, and what produces the entrance into a material manifestation of an entity, a soul:

12 In the beginning was that which set in motion that which is seen in manifested form with the laws governing same. The inability of destroying matter, the ability of each force, each source of power or contact—as it meets in its various forms, produces that which is a manifestation in a particular sphere. This may be seen in those elements used in the various manifested ways of preparing for man, in many ways, those things that bespeak of the laws that govern man's relationship to the first cause, or God.

13 Then, this is the principle:

14 Like begets like. Those things that are positive and negative forces combine to form in a different source, or different manifestation, the combinations of which each element, each first principle manifested, has gained from its associations—

in its activities—that which has been brought to bear by self or that about it, to produce that manifestation.

15 Hence man, the crowning of all manifestations in a material world—a causation world, finds self as the cause and the product of that he (man), with those abilities given, has been able to produce, or demonstrate, or manifest from that he (the soul) has gained, does gain, in the transition, the change, the going toward that (and being of that) from which he came.

16 Periods, times, places: That which is builded, each in its place, each in its time.

17 This is shown to man in the elemental world about him. Man's consciousness of that about him is gained through that he, man, does about the knowledge of that he is, as in relation to that from which he came and towards which he is going.

18 Hence, in man's analysis and understanding of himself, it is as well to know from whence he came as to know whither he is going.

19 Ready for questions.

20 Q What is meant by inequality of experience? Is it a strong argument for reincarnation?

A Considering that which has just been presented, isn't it the same argument?

21 Q Is experience limited to this earth plane?

A As each entity, each soul, in the various consciousnesses, passes from one to another, it—the soul—becomes conscious of that about self in that sphere—to which it, the entity, the soul attains in a materially manifested way or manner.

Hence the entity develops THROUGH the varied spheres of the earth and its solar system, and the companions of varied experiences in that solar system, or spheres of development or activity; as in some ways accredited correctly to the planetary influences in an experience. The entity develops THROUGH those varied spheres.

Hence the sun, the moon, the stars, the position in the heavens or in all of the hosts of the solar systems that the earth occupies—all have their influence in the same manner (this is a very crude illustration, but very demonstrative) that the effect of a large amount of any element would attract a compass. Drawn to! Why? Because of the influence of which the mind element of a soul, an entity, has become conscious!

A soul, an entity, is as real as a physical entity, and is as subject to laws as the physical body as subject to the laws in a material world and the elements thereof!

Does fire burn the soul or the physical body?

Yet, self may cast self into a fire element by doing that the soul knows to be wrong!

What would make a wrong and a right? A comparison of that the soul knows its consciousness to be in accord or contrarywise with, in relation to that which gave it existence. 22 Q Are not transferred memories misappropriated by individuals and considered to be personal experiences?

A Personal experiences have their influence upon the inner soul, while disincarnate entities (that may be earthbound, or that may be heaven-bound) may influence the thought of an entity or a mind.

But, who gives the law to have an element to influence, whether from self or from others? That same as from the beginning. The WILL of the soul that it may be one with the first cause.

In the material, the mental, and the spiritual experience of many souls, many entities, it has been found that there BE those influences that DO have their effect upon the thought of those that would do this or that. Who gives it? Self!

Just as it is when an entity, a body, fills its mind (mentally, materially) with those experiences that bespeak of those things that add to the carnal forces of an experience. Just so does the mind become the builder throughout. And the mental mind, or physical mind, becomes CARNALLY directed!

The mind is the builder ever, whether in the spirit or in the flesh. If one's mind is filled with those things that bespeak of the spirit, that one becomes spiritual-minded.

As we may find in a material world: Envy, strife, selfishness, greediness, avarice, are the children of MAN! Longsuffering, kindness, brotherly love, good deeds, are the children of the spirit of light.

Choose ye (as it has ever been given) whom ye will serve.

This is not beggaring the question! As individuals become abased, or possessed, are their thoughts guided by those in the borderland? Certainly! If allowed to be!

But he that looks within is higher, for the spirit knoweth the Spirit of its Maker—and the children of same are as given. And, "My Spirit beareth witness with thy spirit," saith He that giveth life!

What IS Life? A manifestation of the first cause—God!

23 Q Explain, in the light of reincarnation, the cycle of development towards maturity in individuals.

A As an individual in any experience, in any period, uses that of which it (the soul or entity) is conscious in relation to the laws of the Creative Forces, so does that soul, that entity, develop towards—what? A companionship with the Creative influence!

Hence karma, to those disobeying—by making for self that which would be as the towers of Babel, or as the city of Gomorrah, or as the fleshpots of Egypt, or as the caring for those influences in the experience that satisfy or gratify self without thought of the effect upon that which it has in its own relation to the first cause! Hence to many this becomes as the stumblingblock.

It is as was given by Him, "I am the way. No man approaches the Father but by me." But, does a soul crucify the flesh even as He, when it finds within itself that it must work out its own salvation in a material world, by entering and re-entering that there may be made manifest that consciousness in the soul that would make it a companion with the Creator?

Rather is the law of forgiveness made of effect in thine experience, through Him that would stand in thy stead; for He is the way, that light ever ready to aid when there is the call upon—and the trust of the soul in—that first cause!

Has it not been given that there IS an influence in the mind, the thought of man, from the outside? Then, would those that have lost their way become the guides and both fall in the ditch? or would the soul trust in the Way, and the Light, and seek in that way that there may be shown the light?

What caused the first influences in the earth that brought

selfishness? The desire to be as gods, in that rebellion became the order of the mental forces in the soul; and sin entered.

24 Q What is the strongest argument against reincarnation?

A That there is the law of cause and effect in MATERIAL things. But the strongest argument against reincarnation is also, turned over, the strongest argument for it; as in ANY principle, when reduced to its essence. For the LAW is set—and it happens! though a soul may will itself NEVER to reincarnate, but must burn and burn and burn—or suffer and suffer and suffer! For, the heaven and hell is built by the soul! The companionship in God is being one with Him; and the gift of God is being conscious of being one with Him, yet apart from Him—or one with, yet apart from, the Whole.

25 Q What is the strongest argument for reincarnation?

A Just as given. Just turn it over; or, as we have outlined.

26 We are through for the present.

Appendix

1. The material quoted is from an *A.R.E. Bulletin* article by Gina Cerminara, reprinted in the April 1966 *A.R.E. Journal*.

2. Church, W.H., 1984, *Many Happy Returns*, A.R.E. Press, p. 16. (Reprinted as *The Lives of Edgar Cayce*, A.R.E. Press, 1995, 289 pp.)

3. Cayce, E.E., and H.L. Cayce, 1971, *The Outer Limits of Edgar Cayce's Power*, A.R.E. Press, Virginia Beach, VA 23451, 154 pp.

Subject Index

327

Death ray used on Atlantis 275
Decade of North American Geology (DNAG) study 156
Deluge is not a myth 42
"Directions as to how, where, the elect may be preserved for the replenishing again of the Earth" 42
Displacement of the Earth's spin axis 30
Divine intervention, explanation for earth changes 208
"Division in thy own land" (America) 114
"Domination forces," as Japan's chief feature in 1938, 220
Domination forces in the societies of Germany, Japan, and China 214, 220-222, 281
Dredging to rule out presence of, or to discover, Atlantean remains at Bimini 187
Drug cartels (and mob rule) 121
Dweller on Two Planets, A (book by Phylos) 106

Earth changes
    for 1998 and beyond 180, 235, 262
    interpretations of 191-199
    "no one [knoweth when they will occur] save the Creative Forces" 15, 22
    nuclear war and 224
    outline of, and possible chronology for 191-193
    predictions in the Edgar Cayce readings 1-3
    times and places of 15
    to occur "in the next generation" 247
    "when the changes [will] materially come" [1998-2998 A.D.?] 235
Earthquakes
    as a classic example of a chaotic system 18
    causes of, according to Cayce's readings 65-67, 259
    in the central U.S. 89-99
    in the eastern U. S. 99-106

prediction of lack of, for San Francisco 65
predictions of, are inherently unpredictable 6
"The Terrible Truth" 100
Earth's appearance 10 million years ago 161
"Earth shall be broken up in divers places," at beginning of Earth-change period 17, 45
Earth upheavals during 1936, 12
Earth, what are foundations therof? 209
"Earth will be broken up in many places" 18
"Earth will be broken up in the western portion of America" 257
Eastern Tennessee Seismic Zone 104
Ecuador 141
Egypt 52
Elect may be preserved for the replenishing again of the earth 42
El Salvador 72, 136
Embayments along the U. S. east coast 102
Enormous animals that once roamed the Earth 32
Entrance Point, Bimini 132
Errors in the information obtained by Cayce 250
Eruptions of volcanic calderas 41
Ethiopia 209
Eurasian plate underthrusting Japan 128
Europe (upper portion of) will be changed "in the twinkling of an eye" 58
Exact times for readings-predicted seismotectonic events cannot be given 13
Exotic rock for the carbonate-sediment environment of Bimini 185
Extinction of most of the giant mammals that overran the Earth 33

Final demise of the tundra-steppe biome 33
Final destruction of Atlantis, beginning in 12,700 B.P. and ending in 11,900 B.P. 155

Shinto religion of Japan 222
Sicily 50, 53, 54
Singapore sweatshops 110
"Slime of ages of sea water" covers Atlantean temple off Bimini 182, 184
Soufriere Hills, Montserrat, volcano may be precursor to Pelée eruption 72
Souls given authority concerning the laws of the universe 113
Sources of Edgar Cayce's information 245
South Africa, wheat production after the Earth changes occur 201
South America
    seismic potential 137
    "shall be shaken from the uppermost portion to the end" 137
    tsunami and quake potentials around Lima/Callao 273
South Carolina and Georgia, subsidence of southern portions of 105
Southeast Georgia embayment 103
Southern Hemisphere, inundations by earthquakes 68, 142
South Pacific
    "breakup" of 54
    mantle seismicity of 55
Sphinx 14
Spirit of God must move over the Earth 199
Spitsbergen, recent climate warming in 62
"Star Wars" is back 218
Strategic Defense Initiative (S.D. I. or "Star Wars") 218
Strife between capital and labor 115
Strife is to be "a part of the experience of America . . . because of unbelief" 239
Subduction off the west coast of Honshu and Hokkaido islands 128
Subduction of lithospheric plates under Japan 126-134
Subsidence since mid-Cretaceous times in Japan 131
"Sun will be darkened," at beginning of major Earth changes 17, 42, 45, 257

Suruga Bay, Japan 19

Tambora volcanic caldera 41
Temple Beautiful in Egypt 234
Tennessee, eastern, seismic zone 104
Theory of plate tectonics 146-150
Thera (Thira) 51, 52
Thermal plumes in Earth's mantle 25, 26
"They who have forgotten God must right about face" 225
Tidewater Virginia area as a "safety land" 196
Tierra del Fuego, land to appear off of 19, 188
"Time, and times, and half times" are at an end 15
Times, places, and seasons of Earth changes; who will be given the information as to 18
Tokachi-Oki, Japan, earthquake of 1968 (M 8.2) 128
Tokai, Japan, earthquake of 1854, 132
Tokyo 125
Tonga-Fiji slab 57
Tonga Trench 55-57
Tooele, Utah, chemical weapons depot, largest in America 84
Torrid Zones or areas 22, 39, 40, 43
Torrid Zone volcanoes 41, 43
Towering mountains thrown up in Africa, affecting Nile's course; reading confirmed xii
Transform fault, definition of 261
Tribulation and patience 239
Tropics of Cancer and Capricorn 40
Truman, President H. S., and strife between capital and labor 116
Tsunamis
    definition, and mechanics of generation of 263
    history of, for southern California, Hawaii, etc. 264
Tunesia 53
"Twinkling of an eye," upper portion of Europe will be changed in the 58

Uniformitarianism, doctrine of 248

U. S. Department of Energy's three nuclear weapons laboratories, in western states where Earth is to be "broken up" 217

Universal forces, archangel Halaliel's information obtained from 17

Universal or spiritual laws coming upon Earth 174

Upheaval in 1936, type and extent of 12, 13

Upheavals in the Arctic and Antarctic 18, 39

Upheavals in the interior of the earth 12, 24, 29, 42, 50, 257

Utah 69, 83
    Wasatch fault zone (see also Wasatch Fault) 84

Vancouver Island, Canada 79

Vanuatu, area of explosive activity and "breaking up" in the south Pacific 57

Vesuvius and Pelée 25

Vesuvius, eruptive history 70, 261

Vincennes, Indiana 98

Violence and lust in America, as prelude to Earth changes 110

Volcanic eruptions
    and the "fire" predicted to destroy Earth 42
    due to climate change 38
    in the Torrid Zone 18

Volcanic hazards in Cascadia 81

Wabash Valley fault zone 90

Warren Commission on J.F.K.'s assassination 116

Wasatch fault, Utah 84
    flooding of Salt Lake City by

earthquake on 261
    timing of quakes and "inundations" along 83, 259

Washington (see Pacific Northwest)

West Africa, as an example of a society in criminal anarchy 111

West Antarctica 188

West coast of America, changes predicted for 19

Western portion of America breaking up 18, 65, 257

"What is needed most in the Earth today?" 208

"Who is thy neighbor?" 200, 245

Wisconsin Age glaciation of North America 34

X-ray laser that was to be built at Livermore, California 218

"Ye are to have turmoils . . . etc." (in America) 114

Year (of the Lord) 244

Yellowstone caldera 44

"You attract your life" 226

"You'll get the earth change you deserve" 226

Yucatan 14

Yucca Mountain, Nevada, high-level nuclear waste dump 87

Zaire, America's role in its descent into savagery 271

Zhirov's final destruction of Atlantis 157

Zu, part of, or another name for, the lost continent of Mu in the Pacific 76

Index of Individuals' Names
(With the Exception of Edgar and Gertrude Cayce)

Permissions